Reglazing Modernism

INTERVENTION STRATEGIES FOR 20TH-CENTURY ICONS

Reglazing Modernism

INTERVENTION STRATEGIES FOR 20TH-CENTURY ICONS

Angel Ayón | Uta Pottgiesser | Nathaniel Richards

Birkhäuser
Basel

Contents

Preface

Modernism is the most defining architectural expression of the 20th century. The sheer quantity of Modern-era buildings worldwide is in itself impressive, especially considering that more buildings were built in the 20th century than in all preceding ages combined.[1] From the first late-19th-century examples whose stylistic vocabulary started to depart from Classical architecture to the most recognizable Modern works from the post-WWII period, Modernism transformed the built environment across the globe unlike any other previous period in civilization. The issues affecting these buildings today are increasingly at the forefront of discussions about sustainability, heritage conservation and building sciences, particularly in light of current global challenges like climate change and the need to increase energy efficiency of the built environment.

Suffice it to say that, along with notable quantity and cultural significance, poor performance and limited longevity are also regular hallmarks of Modern buildings. More often than not, these deficiencies involve what is unquestionably one of the most character-defining features of Modern architecture—the exterior glazed enclosures. As a result, the need for interventions to address undesirable conditions at the frames, glass panes and other components of Modern glazed assemblies—referred to in this book generically as reglazing—is ever growing. However, unlike historic materials and assemblies from earlier periods and other architectural expressions, whose intervention criteria has been well established for decades in both the academic and professional fields, there are no proven guidelines validated by time-tested practices that can help to evaluate the appropriateness of past interventions or guide practitioners on current or future projects directed at reglazing Modernism. This book is the first to compile and present critical assessments of a range of intervention approaches to reglazing Modernism. Its goal is to help to fill an informational and analytical void within the profession, and ultimately to help define best practices for intervening on Modern glazed enclosures.

The research presented in this publication is based on the analysis of 20 case studies—nine in the US and eleven in Europe—that together exhibit a wide range of construction typologies and interventions on single-glazed steel frame curtain wall and window wall assemblies. Most of the case study buildings have been retrofitted within the last ten to 15 years in response to one or more driving forces, leading to different intervention approaches—broadly categorized in this book as restoration, rehabilitation or replacement. These categories, the interventions and their motivating factors—such as increasing energy performance or comfort requirements, safety and security concerns, changing functions, and reuse, or material decay, as well as stipulations of historic preservation and heritage conservation—are part of a general discussion in the Intervention Categories section that leads to the presentation of the case studies themselves. In each case study, the book highlights through research and illustrations how the glazed enclosures contribute to the cultural significance of each building, what the conditions were before and after the interventions and offers the authors' opinions on the outcome.

With regard to the selection of the case studies, it was clear early on in the research that reglazing Modernism is far too wide a topic to address in one book. Several strategic choices were made in order to select representative case studies. While these are primarily explained in their own section preceding the case studies, the critical decision to focus mainly on steel frame Modern glazed enclosures is worth noting. There are a wide range of Modern glazed assemblies built with frames made of wood or other metals (like aluminum, bronze or even other types of steel, such as stainless or weathering steel). However, the most commonly encountered assemblies on low-rise, low-density, highly significant Modern buildings are those built with mild steel. Clearly, research on the intervention approaches to other types of Modern glazed assemblies is still needed, and we hope that this book may be only the first of its type in a series that can go on to explore the range of intervention approaches and issues unique to glazed assemblies made out of other materials.

The book concludes with sections summarizing critical findings and pointing to emerging trends derived from the analysis of the case studies. Topics where additional research and development is required are noted, as are high-performance systems that are becoming available in the marketplace as a response from the fenestration industry to the challenges posed by some of the interventions presented in this book. Relevant bibliographic information is also provided so that readers can further explore each case study on their own. While we expect the analysis and criteria included in the book to be of use to both professionals and students, we hope that the graphic information and supporting 3D models will also be of interest to a wider audience.

NOTES

1 As noted by DOCOMOMO co-founder Wessel de Jonge in "Sustainable renewal of the everyday Modern," *Journal of Architectural Conservation 23*, no. 1/2 (2017): pp. 62–105, DOI: 10.1080/13556207.2017.1326555.

Acknowledgments

This research would have been impossible without the generous support of the James Marston Fitch Charitable Foundation, which made a crucial contribution by awarding Angel Ayón the 2015 James Marston Fitch Mid-Career Fellowship. Seri Worden served as the organization's Executive Director during the grant submission and Cristiana Peña since the grant was awarded. Angel Ayón is grateful for the kind letters of support submitted on behalf of his grant application by Nina Rappaport, Kyle Normandin and Pamela Jerome, who has been a mentor since his student days at Columbia University's Historic Preservation program and later on at the office where she was a partner. Ayón is also profoundly grateful for the encouragement and support provided by Theo Prudon and the late Robert Silman throughout the years, and as board members assigned by the Fitch Foundation to advise on the research.

The authors would like to thank DOCOMOMO International, the Association for Preservation Technology International (APTi) and their members for supporting the preparation of the original papers on which this book is based with content and practical input based on publications, symposia and site visits. The authors would also like to acknowledge the leadership of the Facade Tectonics Institute for allowing the preliminary findings of the research in this book to be presented at the 2016 Facade Tectonics World Congress (held 10–11 October 2016, in Los Angeles, California) and including a summary in the published event proceedings.

– Laura Boynton, LEED AP, Assoc. AIA undertook a thorough review and edit of the entire manuscript that went beyond proofreading and grammar review. Her invaluable input and expertise with regard to historic preservation and the renovation of existing buildings helped to ensure the accuracy, completeness and consistency of the ideas and statements included in the book. Without her diligence, the research presented here would not have developed into the form of this book.

– Nicholas Gervasi, formerly with AYON Studio between 2015 and 2017, performed invaluable preliminary archival research on the relevant publications outlining the history, cultural significance and alteration details of most of the case studies included in the research. Sanika Kulkarni from AYON Studio performed additional research for various case studies, including the De La Warr Pavilion. Joseph Gravius from AYON Studio prepared the 3D models for the Hallidie Building, the Van Nelle Factory and Sanatorium Zonnestraal, and assisted with securing copyrights for all the photographs included in the book. Ernesto Antón, formerly with AYON Studio, prepared the 3D models for the De La Warr Pavilion and Lever House. Carlos Rodriguez Estevez from AYON Studio worked on the 3D models for Zeche Zollverein, Fallingwater, Fagus Factory and the Bauhaus Dessau and helped finalizing most of the models prior to rendering. Tomas Fernandez Jer from Atom Studio finalized the models for 14 of the case studies and finalized all of the renders.

– Christine Naumann from the TH OWL in Detmold, Germany assisted with the research of European case studies. Alexis E. (Lexy) Raiford from the University of Florida College of Design, Construction, and Planning assisted with 3D modeling while attending summer school at the TH OWL in Detmold, Germany. Anica Dragutinovic assisted with the research and 3D modeling of the Villa Tugendhat during her internship at the TH OWL. Tavishi Rana and Deepak Singh Dhami, students at TH OWL, prepared the 3D models for the Viipuri Library, Convent of La Tourette and Hardenberg House.

– Fanny Roemer-Lakoubay kindly assisted with translating various quotes from French to English for the case studies on the Van Nelle Factory and the Convent of La Tourette.

– Alexander Severin, a New York City-based architectural photographer, edited most of the photographs in the book, including all taken by Ayón.

Owners and caretakers of the researched properties provided invaluable information on past and/or current conditions and assisted with gaining access to the buildings to evaluate existing conditions. The authors would like to extend a special thanks to the following (shown in order of collaboration date):

– Les Hinmon from AM Universal; Jim Wilcox, Joni Cross and Jon Copaken from Copaken Brooks; Jay Tomlinson from Helix Architecture + Design, Inc.; and R. Jones from SE Associates in Kansas City, Missouri for providing drawings of previous interventions at the Boley Building, which were extremely helpful in determining the original window wall details, as well as of the existing restored cast iron and replacement glass work. Keith Eggener from the University of Oregon, who published an article on the Boley Building that provided valuable insight about past interventions on the building. Michael Sweeney from the State Historical Society of Missouri Research Center in Kansas City also contributed by providing research leads about the Boley Building that, although not ultimately included, was a critical case study researched in preparation for the book.

– Deena Boatman, the Farnsworth House's Operation Manager, who assisted with coordinating Ayón's site visit to the house in Plano, Illinois; and Maurice D. Parrish, the house's Executive Director, who provided valuable clarifications after the site visit regarding previous work and current plans for the house.

– Two unnamed architecture students who generously provided Ayón, his wife and their friend with access to S. R. Crown Hall in Chicago, Illinois after it had been closed to the public for the day.

– Mary Kay Judy who assisted with connecting Ayón to key staff members at Phillip Johnson's Glass House in New Canaan, Connecticut. Group Tour Coordinator Troyce Smith who assisted with the site visit and various questions about the house. Executive Director Greg Sages with assistance from Kate Lichota, Manager of Education and Interpretation and Website Administrator, and Brendan Tobin, Manager of Buildings and Grounds, were also helpful with thorough and thoughtful responses to our various questions.

– Fallingwater docents Ute-Jutta Crooks and Denise Miner who assisted with coordinating Ayón's visit together with Acting Operations Supervisor McKenna Vargo, as well as with Roy Young, Ashley Andrykovitch, Scott Perkins, Clinton Piper, Megan Myers and Roxanne Barnett. Scott W. Perkins, Fallingwater's Director of Preservation, provided valuable clarifications about the replacement glass used as part of the ongoing Window Legacy Fund replacement work.

ACKNOWLEDGMENTS

– Richard W. Southwick, FAIA from Beyer Blinder Belle Architects & Planners LLP (BBB) and Charles W. A. Kramer, AIA, LEED AP from CANY, formerly with BBB, who provided valuable information about the renovation of the TWA Flight Center at JFK Airport.

– Eduardo Faingold, who graciously allowed Ayón, his wife and their daughter to stay at his apartment during their visit to the Yale University Art Gallery in New Haven, Connecticut.

– Winfried Brenne and Franz Jaschke, from Brenne Architekten Gesellschaft von Architekten mbH in Berlin, who welcomed Ayón to their office to discuss their work on the window replacement at the Bauhaus Dessau and also supported Uta Pottgiesser in previous publications. Monika Markgraf, who assisted with coordinating his visit to the Gropius building in Dessau-Roßlau, Germany.

– Nadine Gebauer, Marketing World Heritage Fagus Factory, who assisted with coordinating Ayón's visit to the site in Alfeld, Germany, as well as the staff of the visitor center and store, who were so flexible, welcoming and helpful when he visited with his wife and their daughter.

– Manager Conny Rijbroek and Coordinator Erich Nagel, who assisted Ayón, his wife and their daughter with their visit to Sanatorium Zonnestraal in Hilversum, Netherlands, and Jan Schriefer, who provided a remarkable guided tour of the premises during the site visit.

– Wessel de Jonge graciously met Ayón and his family during their visit to the Van Nelle factory in Rotterdam, Netherlands. Through his work, publications and sharing of knowledge, de Jonge has been an inspiration and mentor to him since they first met in New York City in 2004.

– Emmanuelle Gallo, who co-authored with Vanessa Fernandez two of the articles referenced in the bibliography on the Salvation Army's City of Refuge in Paris, and assisted Ayón in reconnecting with Fernandez, who provided valuable graphic information on the building's original, replacement and existing window and curtain wall assemblies.

– Vanessa Fernandez very kindly provided information about the Convent of La Tourette and City of Refuge in her dissertation and the cited articles, and shared further details and photos and her thoughts on the renovation processes, the state of research and on general heritage preservation with Pottgiesser.

– Tapani Mustonen and Maija Kairamo of the Finnish Committee for the restoration of Viipuri Library at the Alvar Aalto Foundation kindly provided insight into the restoration process of the Library.

– Hans Krabel from Heinrich Böll Architekten in Essen gave his time to explain the different interventions at Zeche Zollverein to Pottgiesser for inclusion in lectures and publications.

– Firma Montag GmbH in Biberach, Germany, who carried out the facade renovation and window replacement work at Hardenberg House, provided several photos and drawings to Pottgiesser for inclusion in lectures and publications.

– Stewart Drew, Director and CEO of the De La Warr Pavilion at Bexhill-on-Sea and Executive Assistant Jo Beattie who assisted Ayón with making arrangements to visit the building. Adam Brown, formerly with John McAslan, who provided valuable information on previous interventions. Sean Albuquerque, Principal at ABQ Studio Chartered Architects in Brighton, UK, also a board member of the De La Warr Pavilion Charitable Trust and steward of the nearby Serge Chermayeff house, who assisted Ayón with understanding the history and development of the site, as well as the scope and extent of previous interventions. And Jasper Goldman, who made arrangements to let Ayón and his family stay at his mother Connie's flat at Marine Court in St. Leonard, Britain's majestic Modernist architectural ocean liner a couple of miles east of Bexhill-on-Sea.

– Iveta Černá, Director of Museum Villa Tugendhat and Barbora Benčiková from the Study and Documentation Centre Villa Tugendhat have provided valuable information and the photos of David Židlický and have always been available for further inquiries.

– Aristónica Ayón Beato and Jesus Torriente Palma assisted Ayón with revising the text during the early stages of preparation for publication. Ayón would like to thank his family for their help, love and support over the years—it means the world to him.

To each and every one of the above, and the numerous individuals and organizations who provided images included in the book: Thank you!

Last, but never least, tremendous gratitude to Ayón's wife, Sarah White-Ayón, who not only came along with their daughter Pilar on several field trips throughout Europe and the US, but also proofread preliminary versions of the research and provided Angel with all the love and support needed through multiple nights and weekends away from his family while preparing this publication.

Introduction

The use of metal-framed glazed enclosures is one of the most distinctive features of Modern architecture. Rooted in a novel approach where the load-bearing building structure became independent from non-load-bearing facades, the early Modern glazed assemblies were the result of new materials and manufacturing processes that came about during the mid- to late-19th century, such as reinforced-concrete, hot-rolled steel profiles and plate glass.[1] The availability and relatively low costs of these materials, in combination with the appeal of their architectural qualities—slenderness, transparency and lightness, for example—resulted in new manufactured exterior enclosure systems such as large ribbon windows, window walls and curtain walls, which were widely used by Modern architects and builders throughout the 20th century. After WWII, the development of new technologies (float glass, brass and aluminum extrusions, for example), coupled with the imperative to rebuild or renew urban centers, led to a sharp increase in the use of exterior glazed assemblies worldwide.

Now, at the beginning of the 21st century, pre- and post-war single-glazed assemblies have been embraced as character-defining features of Modernism. Along with recognition of their cultural significance has come an acknowledgment of their intrinsically poor environmental performance. Efforts to preserve or improve Modern glazed enclosures have lately been at the forefront of real estate and historic preservation debates about the future of these assemblies in the metropolitan marketplace for high-performance buildings. Some have even misguidedly proposed that their poor environmental performance warrants demolition rather than preservation.[2]

Along with the increase in appreciation of Modern architecture as one of the most significant manifestations of the 20th century, there has also been a growing recognition of the need to address the adverse effects of climate change, particularly as related to existing buildings. To date, one of the most influential efforts to fight climate change globally stems from the United Nations Climate Change Conference, better known as COP21, which was held in Paris in the fall of 2015. The conference produced an ambitious international agreement to limit global warming to "well below" 2°C (and below 1.5°C, if possible) higher than pre-industrial levels, which would require net-zero greenhouse gas emissions by the

Former headquarters of the Production Mining Bank in Belo Horizonte, Minas Gerais, Brazil (Oscar Niemeyer, 1953). The original steel frame window walls are in need of repairs as well as functional and performance improvements, 2008.

Planalto Palace, the Presidential office in Brasilia, Brazil (Oscar Niemeyer, 1960). View of the steel frame window walls and the adjacent ramp and roof structures, 2008.

JK Building in Belo Horizonte, Minas Gerais, Brazil (Oscar Niemeyer, 1959). Despite physical decay, the steel frame window walls still display different glass types at the spandrels, vision areas and transoms, 2008.

Annex Building IV of the Chamber of Deputies in Brasilia, Brazil (Oscar Niemeyer, 1977). Almost 20 years after the construction of the first ministry buildings in Brasilia (most with steel frame windows), bronze-colored aluminum profiles, structural tempered glazing and operable aluminum *brise soleils* were used for the north facade of the Annex, reflecting the shift towards aluminum frame construction in the fenestration industry, 2013.

Van Nelle Factory in Rotterdam, Netherlands (Brinkman & Van der Vlugt, 1931). The original steel frame window wall, with its characteristic transparency and modernity, after the rehabilitation work, 2005.

second half of this century.[3] In order to accomplish this goal, energy consumption in the built environment must be significantly reduced. This stark reality has shed light on initiatives like the 2030 Challenge in the US, which aims to make not only new buildings and developments carbon-neutral by 2030, but major renovations as well.[4] Similarly in Europe, the European Building Directive on Energy Performance (EPBD), initially unveiled in 2002 and recently updated in 2018, includes government-mandated requirements to improve energy performance in existing buildings. Defining a pathway towards a "highly efficient and decarbonized building stock by 2050" is a fundamental pillar of the revised EPBD.[5] Accomplishing this goal will require transforming the majority of the existing buildings from being highly energy-inefficient to having nearly net-zero energy use.[6] As a way to monitor progress towards these environmental goals, the EU Building Stock Observatory tracks the energy performance of existing buildings across Europe.[7] They also observe policy implementation in the member states, documenting policy failures and successes across the EU.[8] Comparable standards for the design of energy efficient buildings in the US include: the American Society of Heating, Refrigerating and Air-Conditioning Engineers (ASHRAE) Standard 90.1 (launched in 1975 and updated in 2016); the US Green Building Council LEED program (launched in 1993); the Passive House Institute (founded in Darmstadt, Germany, in 1996); the National Institute of Building Sciences (NIBS) Whole Building Design Guidelines (launched in 1997 and updated in 2002); and other net-zero carbon building initiatives.[9] These non-governmental organizations have voluntary performance requirements that are not as restrictive as the national implementations of their European counterparts.

These strict performance requirements have placed a spotlight on the low performers within the existing building stock and, in particular, those characterized by the extensive use of glazed assemblies. Additionally, many local building codes have mandated their own varying compliance and suitability requirements for safety and energy conservation in historic buildings, exacerbating the need to understand what options are available for intervening in pre- and post-war glazed assemblies.

Consequently, the field of renovating Modern glazed enclosures has become a new branch of knowledge within the areas of heritage conservation and building envelope design, as well as architecture and engineering. To date, the professional discourses have primarily focused on the history, documentation and appreciation of the Modern aesthetic.[10] In addition, there have been discussions on the applicability of well-established historic preservation principles to the conservation of Modern architecture and other conceptual topics regarding authenticity, conservation of historic fabric and original design intent.[11] However, little research has been conducted on technical issues exploring how to reconcile environmental performance and heritage preservation, or on the full range of options available to enhance the appearance of aging metal-framed glazed enclosures, improve their environmental performance and address safety concerns associated with innovative systems with short-lived structural integrity.

Intervention guidelines published in international documents such as the 2017 Madrid-New Delhi Document, published by the International Council on Monuments and Sites (ICOMOS) International Scientific Committee on Twentieth-Century Heritage (ISC20C), focus on documentation, research and management, but do not offer technical guidance.[12] Other documents, such as the intervention principles formulated by the Association for Preservation Technology International (APTi) Technical Committee on Modern Heritage, also focus on documentation, materials, programming and information sharing.[13] They include broad recommendations that are valid as a relevant resource to guide stakeholders on defining the overall approach for intervening on Modern buildings, but offer limited guidance for specific topics such as interventions on Modern glazed assemblies. The lack of specificity on appropriate technical approaches has resulted in either temporary repairs or a tendency to assume that adequate environmental performance should take precedence over historic conservation needs.[14]

Even in the presence of heritage designations that regulate changes to character-defining features, the desire to preserve the original historic fabric and the need to improve performance of the exterior glazed enclosures are often at odds with the conservation of cultural significance in historic buildings, current understanding of building physics and fulfillment of contemporary requirements for building enclosure design and construction. As a result, numerous interventions on Modern metal-framed glazed enclosures involve wholesale removal and replacement with other systems that do not always enhance historic character and cultur-

JK Building in Belo Horizonte, Minas Gerais, Brazil (Oscar Niemeyer, 1959). Ongoing replacement of the original steel frame window walls with a clear anodized aluminum system that has similar sightlines, but different slab edge cover details and finishes, 2008.

al significance. Replacement systems often obliterate original and historic materials, finishes and details, modify design patterns and sightline dimensions, and change subtle but noticeable visible glass properties. Even more concerning is the fact that many of these interventions are often taken as an opportunity to not only upgrade, but also drastically modify a building's original appearance and replace it with one more contemporary.

Yet, the threat of climate change is real and unrelenting, and confronting it is unavoidable. The Second State of the Carbon Cycle Report (SOCCR2), a US federal government-mandated report issued in November 2018, confirmed that urban areas, which occupy only 1–5% of North America, are the primary source of anthropogenic carbon emissions.[15] The same report that cited the built environment (in other words, large infrastructural systems, such as buildings and factories) and the regulations and policies shaping urban form, structure and technology (such as land-use decisions and modes of transportation) are particularly important in curbing urban carbon emissions. The report estimates that the cumulative cost over 35 years of reducing greenhouse gas emissions to meet a 2°C trajectory by 2050 ranges from $1 trillion to $4 trillion (in 2005 USD) in the US. Alternatively, the annual cost of not reducing emissions is conservatively estimated at $170 billion to $206 billion (in 2015 USD) in the US in 2050. With the current slow pace of building renovations reportedly affecting only 0.5–1% of the global building stock annually, and with the impetus to meet the greenhouse gas emission reduction target set forth by the

Sanctuary of Don Bosco in Brasilia (Carlos Alberto Naves, 1963). Steel frame glazing with operable ventilators and various shades of blue stained glass (note the broken glass panes and damaged lead cames), 2006.

Paris Agreement, it is reasonable to assume that the rate and breadth of existing building renovations to achieve energy efficiency will increase exponentially within the next few years.[16]

This increase in the global facades market is expected to reach $337.8 billion (USD) by 2025 from an estimated $179.70 billion (USD) value in 2016.[17] This investment and manufacturing surge in the facade industry, coupled with the aforementioned need to improve the performance of existing buildings, makes the renovation of existing buildings an economic imperative. At the same time, the drive to renovate existing buildings places Modern glazed assemblies at risk of drastic alteration. As stated by DOCOMOMO co-founder Wessel de Jonge, "increasingly stringent requirements have rendered many buildings from the modern era outdated and obsolete—even if they are still performing well according to their original specifications." De Jonge further states that, "it's easy to simply say that they were badly designed, and use that as an excuse to replace them."[18] Thus, there is an urgent need to outline evidence-based recommendations and set forth professional best practices to guide interventions on Modern glazed enclosures. This publication hopes to contribute towards this goal by presenting relevant case studies through a combination of in-depth research and documentation, experience-based critical analysis, as well as compelling and informative construction details.

NOTES

1   The history and development of window walls and curtain walls has been well researched by Ignacio Fernández Solla from Arup's Madrid office, who delved into the topic in a manner unlike anyone else for his unpublished doctoral dissertation at Princeton University in New Jersey (see more about architect Fernández Solla at his blog FacadeConfidential.com). As is well-documented by architectural historians, one of the buildings to first exhibit large glass window walls was Crystal Palace in Hyde Park (Joseph Paxton, 1851). Other similar cast iron buildings with supplemental iron structure followed in New York City (the New York Crystal Palace by Georg Carstensen and Charles Gildemeister, 1852), Munich (Glaspalast by August von Voit, 1854) and other cities (almost all have been demolished, except for a few such as Palacio de Cristal in Madrid's Retiro Park, by Ricardo Velazques, 1887). The rear facades of two smaller-scale buildings designed by Peter Ellis (Oriel Chambers, 1864 and 16 Cook Street, 1866), which still stand in Liverpool, feature prominent glazed enclosures supported by steel frames and cast iron. In the US, cast iron buildings became prominent during the late 1900s, with the facade technology evolving from floor support to steel brackets attached to the edge of the slabs and supporting the cast iron cladding. This is the case at the Temple Court Annex in New York City (James Farnsworth, 1889) and the Hecla Iron Works Building in Brooklyn (Niels Poulson, 1897). The design of large metal-framed glazed facades had significant developments in terms of building size, facade scale and prominence in various European cities such as in Paris at the Au Bon Marche (1898) and in Berlin at the Tietz Kaufhaus store (1898). During the 20th century, this emerging facade technology continued to evolve. One of the most significant examples of the window wall and curtain wall evolution is the Steiff Toy Factory (Giengen 1903). The front facades of the Boley Building in Kansas City, Missouri (Louis S. Curtiss, 1909) and the steel frame curtain wall of the Hallidie Building in San Francisco (William Polk, 1918) ensued. With the exception of the Steiff Toy Factory, all the aforementioned glazed enclosures were designed with the classical vocabulary of the Victorian period. Thanks to their influence, mass-produced unadorned steel frame glazed enclosures went on to become defining elements, enhancing the architecture of Modernism.

2   For further discussion on these debates, see: William Browning et al., *Midcentury (Un) Modern: An Environmental Analysis of the 1958-1973 Manhattan Office Building*, (New York, NY: Terrapin Bright Green LLC, 2014), accessed 24 November 2018, http://www.terrapinbrightgreen.com/wp-content/uploads/2014/03/Midcentury-unModern_Terrapin-Bright-Green-2013e.pdf; James Timberlake, "Should We Save Mid-Century Modern Icons That Hurt The Environment?", *Co-Design*, accessed 24 July 2016, http://www.fastcodesign.com/3054647/should-we-save-mid-century-modern-icons-that-hurt-the-environment; and Angel Ayón and Nina Rappaport, "Greening the Glass Box: A Roundtable Discussion about Sustainability and Preservation," *Mōd* 1 (2014): pp. 17–22.

3   Climate Action UNEP, Sustainable Innovation Forum 2015, "Find out more about COP21," accessed 24 November 2018. http://www.cop21paris.org/about/cop21/.

4   Architecture 2030, "The 2030 Challenge," accessed 24 November 2018, https://architecture2030.org/2030_challenges/2030-challenge/.

5   The European Building Directive on Energy Performance (EPBD) was first published in 2002 and updated in 2010. It is the basis for national regulations in the European Union (EU). See: "Directive (EU) 2018/844 of the European Parliament and of the Council of 30 May 2018 amending Directive 2010/31/EU on the energy performance of buildings and Directive 2012/27/EU on energy efficiency," EUR-Lex, accessed 29 December 2018, https://eur-lex.europa.eu/legal-content/EN/TXT/?toc=OJ%3AL%3A2018%3A156%3ATOC&uri=uriserv%3AOJ.L_.2018.156.01.0075.01.ENG.

6   Buildings Performance Institute Europe (BPIE), *The Concept of the Individual Building Renovation Roadmap. An in-depth case study of four frontrunner projects*," iBRoad, accessed 29 December 2018, p. 1, http://bpie.eu/wp-content/uploads/2018/03/iBRoad-The-Concept-of-the-Individual-Building-Renovation-Roadmap.pdf.

7   EU Building Stock Observatory. "EU Building Stock Observatory," https://ec.europa.eu/energy/en/eubuildings, accessed 29 December 2018.

8   Buildings Performance Institute Europe (BPIE) is an independent not-for-profit think-tank created in 2010 and based in Brussels. Buildings Performance Institute Europe (BPIE), "BPIE Brochure," accessed 29 December 2018, http://bpie.eu/wp-content/uploads/2015/11/BPIE_Brochure.pdf.

9   See: Whole Building Design Guide (WBDG), "Whole Building Design Guide," accessed December 29, 2018, https://www.wbdg.org/resources/whole-building-design; and WBDG, "Building Envelope Design Guide," accessed December 29, 2018, https://www.wbdg.org/guides-specifications/building-envelope-design-guide. Other initiatives include: NGO Architecture 2030, see Architecture 2030, "Existing Buildings: Operational Emissions," accessed 29 December 2018, https://architecture2030.org/existing-buildings-operation/; and "Achieving Zero Framework", see Achieving Zero, "Achieving Zero Framework," accessed 29 December 2018, http://achieving-zero.org/.

10 This is reflected in DOCOMOMO'S Eindhoven-Seoul Statement: DOCOMOMO International, "Eindhoven-Seoul Statement 2014," accessed 24 November 2018, http://www.docomomo.com/eindhoven.

11 For further discussion, see: Susan Macdonald and Gail Ostergren, eds., *Conserving Twentieth-Century Built Heritage: A Bibliography*, Los Angeles: Getty Conservation Institute, 2013, accessed 24 November 2018, http://hdl.handle.net/10020/gci_pubs/twentieth_century_built_heritage; and Angel Ayón, "Historic Fabric vs. Design Intent: Authenticity and Conservation of Modern Architecture at Frank Lloyd Wright's Solomon R. Guggenheim Museum," *Journal of Architectural Conservation* 15, no. 3 (2009): pp. 41–58.

12 ICOMOS International Scientific Committee on Twentieth-Century Heritage (ISC20C), *Madrid-New Delhi Document: Approaches for the Conservation of Twentieth-Century Architectural Heritage*, 2017, accessed 24 November 2018, http://www.icomos-isc20c.org/pdf/madrid-new-delhi-document-2017.pdf.

13 David Fixler, "Toward APT Consensus Principles for Practice on Renewing Modernism," *APT Bulletin* 48, no. 2/3, Special Issue on Modernism (2017): pp. 6–8.

14 For in-depth discussion on this issue, see: Pamela Jerome and Angel Ayón, "Can the 1960s Single Glazed Curtain Wall Be Saved?" *APT Bulletin* 45, no. 4 (2014): pp. 13–19; and Browning et al., *Midcentury (Un)Modern: An Environmental Analysis of the 1958–1973 Manhattan Office Building*.

15 N. Cavallaro et al., *Second State of the Carbon Cycle Report (SOCCR2): A Sustained Assessment Report* (Washington, DC: U.S. Global Change Research Program, 2018), Report In Brief pp. 17 and 22, https://doi.org/10.7930/SOCCR2.2018.

16 Architecture 2030, "Existing Buildings: Operational Emissions," accessed 24 November 2018, https://architecture2030.org/existing-buildings-operation/.

17 Research and Markets, "Global Facades Market Analysis and Segment Forecasts, 2014-2017 & Forecasts to 2025: Key Players are Ametek Inc., Franklin Electric Co., Asmo Co. Ltd., and ABB," accessed 24 November 2017, https://www.prnewswire.com/news-releases/global-facades-market-analysis-and-segment-forecasts-2014-2017--forecasts-to-2025-key-players-are-ametek-inc-franklin-electric-co-asmo-co-ltd-and-abb-300592638.html.

18 Wessel de Jonge, "Sustainable renewal of the everyday Modern." *Journal of Architectural Conservation* 23, no. 1/2 (2017): pp. 62–105. DOI: 10.1080/13556207.2017.1326555.

Reglazing Modernism and Historic Preservation

The interest and need for knowledge on reglazing Modernism as referenced in this publication has been a topic of discussion for at least 20 years, particularly through the International Working Party for the Documentation and Conservation of Buildings, Sites, and Neighborhoods of the Modern Movement known since 1990 as DOCOMOMO.[1] The work of this international organization is primarily based on documentation and advocacy efforts by national and local chapters, as well as on contributions from the various International Specialist Committees (ISCs) established to address particular subjects, such as technology, urbanism, interior design and education, among others.[2] In 1996, a seminar in Eindhoven organized by the DOCOMOMO International Specialist Committee of Technology (ISC-T) discussed the issue of curtain wall refurbishment as a "challenge to manage". In his introduction, DOCOMOMO co-founder Wessel de Jonge referred to the "increasing requirements" that pre- and post-war buildings with light facade constructions had to comply with, and also noted that "the renovation market will develop from quite a marginal niche into a much more significant volume in turnover since the number of renovation jobs will grow exponentially over the coming years."[3] This has certainly happened in the European building sector, where renovation makes up approximately 70-80% of the market.[4] However, the level of improvement required to reach the performance values specified by European guidelines for energy efficiency has not increased significantly.[5] Individual investors and residential building owners are lagging behind commercial and industrial properties, which are then faced with demands to contribute even more. In this context, Modern Movement heritage buildings are often the first to be renovated and thus become fertile ground to test and verify different intervention approaches.

Several publications have addressed the topic of conservation of Modern steel frame assemblies. In 1999, DOCOMOMO International dedicated Journal 20 to "Windows to the Future", summarizing different approaches to interventions on steel frame assemblies, including studies on windows by English Heritage, PVC-U Replacement Windows in Tallinn, as well as replacement systems at Copenhagen's White Meat Town (1932–1934) and at Rietveld's School of Art, Arnhem (1958–1963).[6] A year later, a DOCOMOMO preservation technology dossier on "Reframing the

Moderns" summarized the history and development of metal-framed glazed enclosures, outlined strategies and policies for interventions, and presented ten case studies assessing Modern glazed enclosures of various ages and locations in both the US and Europe.[7]

In the US, the topic has also received increasing attention over the years in both publications and symposia. The Association for Preservation Technology International (APTi) is a North American cross-disciplinary membership organization dedicated to promoting the best technology for conserving historic structures and their settings. Their triannual publication, the *APT Bulletin*, has included several articles on the conservation of Modern curtain walls. In 2001, a full issue of the *APT Bulletin* was dedicated to the preservation of the 20th-century curtain wall. Article topics ranged from curtain wall origins and performance fundamentals to discussion of preservation issues, conflicts and challenges and authenticity considerations when altering these systems.[8]

The 2004 DOCOMOMO International Conference in New York City included a technology session on the "20th-Century Metal and Glass Curtain Wall." During the session, several case studies and a comprehensive draft timeline on the history of curtain wall testing and evolution were presented.[9]

The Skyscraper Museum in New York City and DOCOMOMO-New York/Tri-State held two panel discussions about Modern buildings in midtown Manhattan, including one in 2007 about preserving midtown Modernism and another one in 2008 about how to make them more sustainable.[10] While the buildings in midtown Manhattan seldom have steel frame Modern glazed assemblies (most date from the post-WWII period and were primarily built with aluminum frames), the presentations had a unique emphasis on the glazed facades and how to upgrade them.[11] In his seminal 2008 publication, *Preservation of Modern Architecture*, Theo Prudon emphasized in regards to various relevant case studies worldwide that knowledge about the technological challenges related to the renovation of Modern glazed assemblies, as well as their solutions, is one crucial point to start with a broader refurbishment.[12]

Other relevant discussions on the renovation of Modern glazed facades took place during the 10th International DOCOMOMO Conference which convened in Rotterdam in September 2008. While discussing the challenges posed by the legacy of the Modern Movement, the topics presented at a session called "Technology, Progress and Sustainability" included the conservation and adaptive reuse of Modern building facades, insulating issues related to single-glazing, natural ventilation of Modern buildings, a comparative analysis of Modern steel frame glass walls from Germany and Brazil, as well as an analysis of the synergies between sustainability and conservation of the Modern Movement.[13] The European Facade Network (efn) was established a short time later in 2009. With five university partners, it seeks to advance and promote facade design and engineering through education and research that focuses on innovations for new building envelopes and for rehabilitation and reuse.[14] The *APT Bulletin's* "Special Issue On Modern Heritage" was published in 2011, and included papers summarizing interventions on glazed assemblies at the UNESCO Headquarters (Bernard Zehrfuss, Marcel Breuer and Pier Luigi Nervi, 1958), the Yale University Art and Architecture Building (Paul Rudolph, 1963) and the Solomon R. Guggenheim Museum (Frank Lloyd Wright, 1959).[15]

In 2013, a roundtable discussion on sustainability and preservation with building envelope and preservation practitioners, including Gordon Smith, Robert Heintges, Israel Berger and Pamela Jerome, was convened by DOCOMOMO-New York/Tri-State during the Association for Preservation Technology International's "Preserving the Metropolis" conference in New York City. The roundtable presenters discussed the challenges of preserving Modern buildings with glazed facades and started to define best practices for developing a strategy to safeguard the most significant of these buildings.[16] Also in 2013, the Getty Conservation Institute (GCI) published a bibliographic compendium entitled *Conserving Twentieth-Century Built Heritage* that includes numerous entries, including references to the aforementioned symposia, as well as to other interventions on Modern glazed assemblies around the world.[17]

The following year, in 2014, the Twentieth Century Heritage International Scientific Committee (ISC20C) of the International Council of Monuments and Sites (ICOMOS) published the international standard *Approaches for the Conservation of the Twentieth Century Architectural Heritage.* Known as the Madrid Document, and extended to include urban areas and landscapes after the 2017 ICOMOS General Assembly in Delhi, this reference text was among the first to outline the approach and principles "that should be applied to managing and interpreting twentieth-century sites and places."[18] Also aiming to establish practice guidelines, the APT Twentieth Century-Modern Heritage Committee (APTi TC-MH) organized the "Renewing Modernism: Emerging Principles for Practice" symposium as a conclusion to the 2015 APT International Conference: Convergence of People and Places—Diverse Technologies and Practices. The symposium included various presentations on European and North American buildings, with many focusing on the rehabilitation of Modern metal frame glazed facades. The outcome of this event was a set of "consensus principles for practice" aimed at guiding interventions on Modern buildings that were later included in another issue of the *APT Bulletin* focused on Modernism.[19] These international documents and consensus principles are relevant and necessary to build the conceptual framework required to guide and evaluate the decision-making process for intervening on Modern buildings and sites. They are also, however, broad and lack technical specificity. Thus, they have a limited impact on the myriad of architecture, engineering, building science and facade consulting practitioners and other stakeholders tasked with crafting facade renovation solutions that are historically appropriate, cost effective, environmentally viable, respectful of the past and informed by current building regulations and facade technology. Another emerging forum that is bringing together all of these stakeholders is the biennial world congress organized by the Facade Tectonics Institute (FTI). The Renovation/Preservation track in the 2016 event, as well as the Heritage Facade track in the 2018 event, included technical presentations focused on interventions to Modern glazed assemblies of historical and architectural significance.[20]

NOTES

1   Theodore H. M. Prudon, *Preservation of Modern Architecture* (Hoboken, N.J.: John Wiley & Sons, 2008), p. 11. For more on the history and development of DOCOMOMO, see *DOCOMOMO Journal 27: The History of DOCOMOMO*, edited by Wessel de Jonge et al, no. 27 (June 2002), entire issue. For an overview of the history of conserving Modern architecture, see Chapter 1 of Prudon, *Preservation of Modern Architecture*, pp. 1–22.

2   Dossiers dealing with particular aspects of building technology and materials as they affect the construction and restoration of Modern architecture have been published by ISC Technology editor P. Tournikiotis.

3   Wessel de Jonge and Arjan Doolaar, eds., "Curtain Wall Refurbishment. A Challenge to Manage," *Proceedings of the International DOCOMOMO Seminar, Eindhoven University of Technology, the Netherlands, January 25 1996*.

4   It was not possible to locate reliable data for the US at the time of printing.

5   Martin Gornig, Christian Kaiser and Claus Michelsen, "German Construction Industry: Refurbishment Lacks Momentum, New Residential Construction Gets Second Wind," *DIW Economic Bulletin* 49 (2 December 2015): p. 641, accessed 24 November 2018, https://www.diw.de/documents/publikationen/73/diw_01.c.521558.de/diw_econ_bull_2015-49-1.pdf.

6   See *DOCOMOMO Journal* 20, "Windows to the Future" (January 1999), entire issue.

7   See *DOCOMOMO Preservation Technology Dossier 3*, "Reframing the Moderns, Substitute Windows and Glass" (April 2000), entire issue.

8   See *APT Bulletin: The Journal of Preservation Technology* 32, no. 1, "Curtain Walls" (2001), entire issue.

9   See *DOCOMOMO Preservation Technology Dossier 8*, "Restoring Postwar Heritage: Selections from the 2004 DOCOMOMO US Technology Seminar" (August 2008), entire issue. In reference to the timeline, see: Daniel J. Lemieux and Martina T. Driscoll, "The History of the 20th Century Metal and Glass Curtain Wall, and Evolution of Curtain Wall Standardization and Performance Testing," Handout distributed during the 8th International DOCOMOMO Conference, DOCOMOMO US Technology Seminar, New York City, 26 September–2 October, 2004.

10 The Skyscraper Museum and DOCOMOMO-New York/Tri-State, "RE:NY Recycle | Retrofit | Reinvent the City," Panel discussion, filmed 5 February 2008, accessed 24 November 2018, https://www.skyscraper.org/PROGRAMS/ReNY/reny01_01.php.

11 Nina Rappaport and Erik Sigge, "The Midtown Manhattan Project," *DOCOMOMO Journal* 31 (September 2004): p. 113.

12 Prudon, *Preservation of Modern Architecture*, pp. 126–128.

13 Dirk Van Den Heuvel et al., eds,. *The Challenge of Change: Dealing with the Legacy of the Modern Movement*, Proceedings of the 10th International DOCOMOMO Conference. Amsterdam: IOS Press, 2008.

14 European Facade Network (efn), "About," accessed 24 November 2018, http://facades.ning.com/page/about.

15 See *APT Bulletin: The Journal of Preservation Technology* 42, no. 2/3, "Special Issue On Modern Heritage" (2011), entire issue.

16 See Angel Ayón and Nina Rappaport, "Greening the Glass Box: A Roundtable Discussion about Sustainability and Preservation," *Mōd* 1 (2014): pp. 17–22.

17 See Macdonald and Ostergren, *Conserving Twentieth-Century Built Heritage*.

18 ICOMOS ISC20C, *Madrid-New Delhi Document*, p. 2.

19 See Fixler, "Toward APT Consensus," pp. 6–8.

20 The proceedings of both the 2016 and 2018 Facade Tectonics Conferences are available online for download, accessed 24 November 2018, https://facadetectonics.org/publications/.

Case Studies

Introductory Notes

SELECTION CRITERIA

For the sake of allowing for a balanced comparative analysis, most of the case studies included in this book are buildings with steel-framed enclosures whose national and/or international significance has been long established through architectural history or heritage designations. Thus, modest buildings with known reglazing interventions whose cultural significance is mostly local and does not reach the national or international level were excluded. That includes, among others, the Boley building in Kansas City, Missouri. Designed by Canadian-born American architect Louis Singleton Curtiss (1865–1924) and built in 1909, the building has been referred to as having America's first curtain wall.[1] However, during the preparatory research for this book, when the documentation on previous interventions was located and reviewed by the authors, it was revealed that this claim is incorrect—the building features a window wall instead of a curtain wall. This finding demoted the significance attributed to the building and was the basis for not including it as a case study, even though a full analysis had already been completed.

While the case studies selected for this book consist of North American and European Modern buildings only, this limitation is by no means an attempt to exclude other relevant case studies in places such as Central and South America, Asia, Africa, Australia, Oceania or elsewhere. Although some may be unknown to the authors, there are many more case studies from other regions that deserve to be included in future editions or in other publications.

In the end, 20 case studies made the cut. Nine are in the US—six in the northeast, two in the midwest, and one on the west coast. Eleven are in Europe—four in Germany, two in the Netherlands, two in France, one in the UK, one in the Czech Republic and one that was built in Finland but is now in Russia. This last case refers to the Viipuri Library (now known as the Vyborg Library), which has the unusual circumstance of having been built in the Finnish city of Viipuri, which was then annexed by the former USSR during WWII and renamed Vyborg. Together, the case studies selected exhibit a wide range of construction typologies and interventions on single-glazed steel frame curtain wall and window wall assemblies.

It should be noted that references to each project's overall construction cost or the cost of the work related to the exterior steel frame glazed assemblies, although known for selected case studies, were excluded. This exclusion prevented the disclosure of confidential or unpublished information. It also enabled the reader to avoid uninformed evaluations related to project cost, as this important project component is dependent on conditions unique to each project region, such as the local labor market and availability of materials. It is also worth noting that case studies of steel frame glazed assemblies from other cultural expressions following the Modern period, such as Post-Modernism, were excluded. Likewise, interventions to frameless, contemporary all-glass or point-supported enclosures were also intentionally excluded. These exclusions allowed the authors to concentrate on conditions and interventions on buildings from a single period.

CASE STUDY PRESENTATION FORMAT

The case studies are organized in groups by their intervention approach—restoration, rehabilitation or replacement—which are further discussed below. Each of these intervention categories is illustrated by projects of varying cultural significance, technical complexity, cost and outcome. The case studies selected showcase a number of intervention approaches on Modern glazed enclosures using historic images, technical details and photographs that show conditions before and after the interventions were performed.

Each presentation follows a consistent format devised to clearly convey critical information for each building. First, under the building name heading, the building location is listed, followed in parentheses by the original architect's name and the year the original construction was completed. A brief narrative then provides a description of the building and a summary of its cultural and/or architectural significance with an emphasis on how the steel frame exterior enclosures help to define said significance. Subsequently, the Condition Prior to Intervention and Intervention sections summarize changes to the enclosure over the years (if applicable) and outline the existing conditions and driving factors motivating the most relevant reglazing intervention. The Comments section concludes each case study with the authors' assessment of the impact of the restoration, rehabilitation or replacement work on the values defining the significance of each building and its glazed enclosure. The comments evaluate each case study in terms of how the intervention helped to preserve culturally significant features and how it adhered to the current understanding of building physics and contemporary requirements for building envelope performance.

The evaluation of each case study is informed by the authors' familiarity with each building, including personal observations during site visits and a review of published sources. Another critical source of information for the observations presented on each case study were project presentations and direct discussions with the architects, engineers and/or building envelope consultants responsible for the latest alterations. These exchanges with colleagues allowed for a better understanding of each project's conservation requirements, budget and programmatic constraints. In more than one case study, these exchanges allowed the authors to obtain or clarify information unavailable in publications. However, the evaluations included in this book represent the opinions of the authors and not necessarily those of the various building owners, project team members and other stakeholders contacted by the authors while preparing the material presented here.

In addition to the narrative information, a graphic inset includes a timeline of notable dates related to the building design, construction, alterations and heritage designations, and also provides information on the original and intervening teams (as available).

The technical details from before and after the interventions in each case study are illustrated with photographs and rendered 3D details prepared specifically for this publication. For case studies where the reglazing intervention resulted in changes to the original details, 3D detail sections are provided to show the condition before and after alteration. In those where the reglazing intervention resulted in no relevant changes to the original detail and configuration of the glazed enclosure, a slab-to-slab wall section is rendered in 3D. These details aim to visualize both the nature and configuration of the exterior glazed enclosures and the content of the technical information outlined in the narrative for each case study.

Intervention Categories

Alterations to Modern glazed enclosures are often guided by design professionals who, while trained and experienced with facade technology, building physics and related disciplines, are not always familiar with long-established built heritage conservation theory and practice, or well-versed in the terminology underlying such disciplines. As a result, trade publications and most of the bibliographic resources reviewed in preparation for this publication refer to interventions on Modern glazed enclosures using a variety of terms such as repairs, renovation, refurbishment, renewal and upgrading, as well as recladding, overcladding and retrofitting. From the perspective of built heritage conservation professionals, these terms are imprecise and open to interpretation. In an effort to adhere to established historic preservation practice in the US, the case studies presented in the next pages are organized according to three main categories—restoration, rehabilitation and replacement. The first two categories are widely used terms that refer to specific interventions on historic buildings, regardless of era or significance. These two terms are defined in the *US Secretary of the Interior's Standards for the Treatment of Historic Properties.* The *Standards*, together with the accompanying *Guidelines for Preserving, Rehabilitating, Restoring and Reconstructing Historic Buildings*, provide a set of concepts and prescriptive recommendations that apply to the maintenance, repair and replacement of historic materials, and to the design of additions and alterations at historic properties. These documents provide the regulatory standard for projects receiving federal and other government funding, but also offer general guidance for work on historic buildings in the US.[2] They were last revised in 2017 to include definitions and recommended treatments for Modern features such as curtain walls, as well as to promulgate sustainability requirements and to address present-day challenges such as resilience to natural hazards.

The definitions of restoration and rehabilitation used for the intervention categories in this book follow closely those outlined by the *Standards.* As such, restoration is defined as "the act or process of accurately depicting the form, features, and character of a property as it appeared at a particular period of time by means of the removal of features from other periods in its history and reconstruction of missing features from the restoration period."[3] The *Standards* define rehabilitation as "the act or process of making possible a compatible use for a property through repair, alterations, and additions while preserving those portions or features which convey its historical, cultural, or architectural values."[4]

The third main intervention category used in this book—replacement—is more self-explanatory. This is a treatment deemed appropriate when the physical condition of distinctive materials and features is severely deteriorated or beyond its life expectancy and requires removal and substitution with in-kind or look-alike materials. The *Standards* refer to limited replacement as an appropriate component of both restoration and rehabilitation treatments. Extensive replacement, with its implicit loss of historic fabric, however, is discouraged by the *Standards* for all treatments.

In general, these definitions are too broad to capture the full range of conditions that afflict Modern steel frame glazed enclosures. For these facade components, it has long been established that interventions are not only triggered by physical decay. They are also the result of deficient structural capacity; poor thermal performance and the resulting high energy use and poor user comfort; lack of water- and air-tightness and the resulting leaks; air exfiltration, infiltration and condensation; poor acoustic isolation and fire protection; and extensive exposure to daylight and the resulting solar heat gain and glare, just to mention a few.[5] Consequently, the use of well-established historic preservation terms requires fine-tuning to make them applicable to reglazing interventions. The restoration, rehabilitation and replacement approaches outlined in the case studies presented on the following pages are better defined as summarized below.

RESTORATION
STEEL REPAIRS AND SINGLE-PANE GLASS REPLACEMENT

Several case studies represent restoration, where the use and significance of the building or stated interest in retaining the original architectural character led to the repair and maintenance approach. Restoration is a limited intervention that focuses on repair of the steel frames. As part of this approach, severely damaged components, usually affected by corrosion, are cut out and locally replaced in-kind. This restorative approach is illustrated in Fallingwater in Mill Run, PA, USA (Frank Lloyd Wright, 1936), the Vyborg (Viipuri) Library in Russia (Alvar Aalto, 1935) and the Hallidie Building in San Francisco, CA, USA (Willis Polk, 1918). In other cases, the steel frames are mostly left untouched and only the glass panes are replaced, as at the Glass House in New Canaan, CT, USA (Phillip Johnson, 1949), the Farnsworth House in Plano, IL, USA (Ludwig Mies van der Rohe, 1951), the TWA Flight Center at JFK Airport in Queens, NY, USA (Eero Saarinen, 1962), and the Villa Tugendhat in Brno, Czech Republic (Ludwig Mies van der Rohe, 1930).

REHABILITATION
GLASS REPLACEMENT WITH IGU OR ADDITION OF SECONDARY GLAZING

Rehabilitation is often implemented when the original glass panes have been lost, are not commercially available, or where it is not financially feasible or desirable to replicate them. Further, new uses proposed for the building or its continued use according to contemporary performance requirements have led to solutions where the steel frames are repaired or replaced and new insulated glass units (IGUs) or additional layers are added to improve energy performance and comfort aspects. This rehabilitation approach is illustrated by the Hardenberg House in Berlin (Paul Schwebes, 1956), the Van Nelle Factory in Rotterdam, Netherlands (Brinkman & Van der Vlugt, 1931), the Zeche Zollverein complex in Essen, Germany (Schupp and Kremmer, 1932/1961), and the De La Warr Pavilion in Bexhill-on-Sea, UK (Erich Mendelsohn and Serge Chermayeff, 1935).

REPLACEMENT
WITH (NON-)THERMALLY BROKEN STEEL OR ALUMINUM FRAMES

Replacement is described in some examples where, despite the significance of the building and its original architectural character, the original steel frames and glazing have been replaced by new assemblies. In general, this approach is aimed at retaining the original appearance as much as possible while improving the building's overall performance. Energy conservation and user comfort are typically the main reasons behind replacement in the European case studies. For the US case studies illustrating the replacement approach, this more invasive intervention is directed at enhancing building envelope performance and responding to safety requirements. This is illustrated by the case studies of S. R. Crown Hall in Chicago, IL, USA (Ludwig Mies van der Rohe, 1956), the Solomon R. Guggenheim Museum, New York, NY, USA (Frank Lloyd Wright, 1959), and Lever House, New York, NY, USA (SOM, 1952), and the Yale University Art Gallery in New Haven, CT, USA (Louis Kahn, 1953). The same approach is applied in the European case studies of Fagus Factory in Alfeld, Germany (Walter Gropius and Adolf Meyer, 1912/1925), Sanatorium Zonnestraal in Hilversum, Netherlands (Jan Duiker, 1928–1931), the Bauhaus workshop, office and residential studio buildings in Dessau-Roßlau, Germany (Walter Gropius, 1926), the Convent of La Tourette in Éveux, France (Le Corbusier, 1960), and City of Refuge in Paris, France (Le Corbusier and Pierre Jeanneret, 1933/1953).

The following table summarizes key information for all 20 case studies, helping to visualize the different approaches for each project. To this end, the table also organizes the case studies into groups under the intervention categories of restoration, rehabilitation and replacement. Each case study entry includes the general building information, the main intervention category and the main drivers of each intervention, as well as a quick reference to what the enclosure consisted of before and after the intervention.

NOTES

1   Keith Eggener, "The Uses of Daylight: Louis S. Curtiss, the Boley Building, and the Invention of the Glass Curtain Wall," *Places Journal* (May 2012), doi.org/10.22269/120514.

2   Anne E. Grimmer and Kay D. Weeks, *The Secretary of the Interior's Standards for the Treatment of Historic Properties with Guidelines for Preserving, Rehabilitating, Restoring & Reconstructing Historic Buildings*, (Washington, D.C.: U.S. Department of the Interior, National Park Service, Technical Preservation Services, rev. ed., 2017), pp. 2–3.

3   Ibid., p. 163.

4   Ibid., p. 75.

5   Jacques Mertens, "Curtain walls as a system of building physics—A perspective for refurbishment," in *Curtain Wall Refurbishment: A Challenge to Manage*, eds. Wessel de Jonge and Arjan Doolar (Eindhoven: DOCOMOMO International, Eindhoven University of Technology, 1997), pp. 35–40.

Data Synthesis

	BUILDING NAME (Completion year)	CITY	COUNTRY	Energy conservation	Safety/security	Materials decay	User comfort	Historic preservation / heritage conservation	BEFORE [Frame/glass]	AFTER [Frame/glass] (Completion year)
				MOTIVATING FACTORS FOR INTERVENTION						REGLAZING MODERNISM INTERVENTION
RESTORATION										
I-1	Villa Tugendhat (1930) Window walls Entrance staircase	Brno	Czech Republic			●		●	ST / SG ST / SG ST / SG	ST / SG (1945) ST / SG (2012) ST / SG (2012)
	Glass House (1949)	New Canaan, CT	USA			●		●	ST / SG	ST / SG (as needed)
I-2	Hallidie Building (1918)	San Francisco, CA	USA			●		●	ST / SG	ST-SR / SG-LG (2014)
	Viipuri Library (1935) Window wall (stairs, lobby) Skylights (lecture and lending hall)	Vyborg	Russia	●	●		●	●	ST / SG-IGS WD / SG	ST / SG-IGS (1995, 2010) WD / SG+2G-LG (2005)
	Fallingwater (1937)	Mill Run, PA	USA		●			●	ST / SG	ST / SG-SF (1989, 1993) ST / SF-LG (ongoing)
	Farnsworth House (1951)	Plano, IL	USA		●			●	ST / SG	ST / SG (1996, 2013, TBD)
	TWA Flight Center (1962)	New York, NY	USA			●		●	ST / SG	ST / SG-LG (2012-2019)
REHABILITATION										
II-1	Zeche Zollverein (1932/1961) Halls 2 and 5 Halls 7 and 9 Hall 9 (selected locations) Halls 6 and 10	Essen	Germany	●			●	●	ST / SG ST / SG ST / SG ST / SG	ST / SG (1992, 1993) ST / IGU (1996, 1997) ST-TB / IGU (1996) ST / 2G-IGU (1992, 1993)
II-2	Van Nelle Factory (1931)	Rotterdam	Netherlands	●			●	●	ST / SG	ST / SG+ 2G-IGU (2004)
	Hardenberg House (1956) Street facades Courtyard facades	Berlin	Germany	●		●	●	●	ST / SG-IGS ST / SG-IGS	ST / SG-IGS (2004) AF-TB / IGU (2004)
II-3	De La Warr Pavilion (1935)	Bexhill-on-Sea	United Kingdom		●	●		●	ST / SG	ST-SR / SG (unknown)

INTERVENTION CATEGORY (Only main one is shown)

ABBREVIATIONS

Frames: ST: steel frame; WD: wooden frame; AL: aluminum frame; BF: brass frame; PF: polyurethane frame; SF: silicone frame; CF: concrete frame; TB: thermally broken; SR: structural reinforcement

Glass: SG: single glazing; IGU: insulated glass units; 2G: secondary glazing; SF: surface-mounted film; LG: laminated glass; IGS: interior glazed shutters

INTERVENTION CATEGORY (Only main one is shown)	BUILDING NAME (Completion year)	CITY	COUNTRY	Energy conservation	Safety/security	Materials decay	User comfort	Historic preservation / heritage conservation	BEFORE [Frame/glass]	AFTER [Frame/glass] (Completion year)
				MOTIVATING FACTORS FOR INTERVENTION					REGLAZING MODERNISM INTERVENTION	
	REPLACEMENT									
III-1	Lever House (1952)	New York, NY	USA			●		●	ST / SG	ST / SG-LG (2001)
	S. R. Crown Hall (1956)	Chicago, IL	USA		●	●		●	ST / SG	ST / SG-LG (2006)
	Convent of La Tourette (1960)	Éveux	France		●	●		●	CF BF-PF / SG	CF-AL-SF / SG (1993) CF-BF-PF / SG-LG (2013)

III-2	Fagus Factory (1912/1925)
	    Office building
	    Office building corners, etc.
		Alfeld	Germany	●		●	●	●	
									ST / SG
									ST / SG
										
										ST / IGU (ca. 1990)
										ST / SG (ca. 1990)

	Sanatorium Zonnestraal (1928–1931)	Hilversum	Netherlands			●		●	ST / SG	ST / SG/IGU (2006)

	City of Refuge (1933/1953)
	    Penthouse
	    Main facades
		Paris	France	●			●	●	
									ST / SG
									ST / SG
									ST / SG
									WD / SG
										
										ST / SG (1953)
										ST / IGU (2015)
										WD / SG (1953)
										WD / IGU (2015)

III-3	Bauhaus Dessau (1926)
	    Workshop
	    North facade
	    Residential studios
		Dessau-Roßlau	Germany	●			●	●	
									ST / SG
									ST / SG
									AL / SG
									ST / SG
									AL / SG
										
										AL / SG (1976)
										AL / SG (1976)
										ST-TB / IGU (2015)
										AL / SG (1976)
										ST-TB / IGU (2015)

	Solomon R. Guggenheim Museum (1959)	New York, NY	USA	●				●	ST / SG	ST-TB / IGU (2008)
III-4	Yale University Art Gallery (1953)	New Haven, CT	USA	●		●		●	ST / SG	AL-TB / IGU (2006)

INTERVENTION CATEGORIES

I. Restoration
1. Frame repair and single-pane glass
2. Frame repair and single-pane laminated glass

II. Rehabilitation
1. Frame repair and insulating glass units (IGU)
2. Frame repair and single-pane glass plus secondary glazing
3. Frame repair with structural reinforcement and single-pane glass

III. Replacement
1. Non-thermally broken steel frames and single-pane glass
2. Non-thermally broken steel frames and insulated glass units (IGU)
3. Thermally broken steel frames and insulated glass units (IGU)
4. Thermally broken aluminum frames and insulated glass units (IGU)

Classification
Overview

RESTORATION INTERVENTION TYPES

ST / SG
Frame repair +
single-pane glass

ST / SG-LG
Frame repair +
single-pane laminated
glass

ST / SG-SF
Frame repair +
single-pane glass +
surface-mounted film

Villa Tugendhat, 1930

Hallidie Building, 1918

Fallingwater, 1937

Zeche Zollverein,
1932/1961

Villa Tugendhat, 1930

Farnsworth House, 1951

Viipuri Library, 1935

Viipuri Library, 1935

TWA Flight Center, 1962

Glass House, 1949

Fallingwater, 1937

Farnsworth House, 1951

TWA Flight Center, 1962

TWA Flight Center, 1962

REHABILITATION INTERVENTION TYPES

ST / IGU
Frame repair + IGUs

Sanatorium Zonnestraal, 1928–1931

Zeche Zollverein, 1932/1961

ST-SR / SG+
Frame repair with structural reinforcement + single-pane glass

Hallidie Building, 1918

De La Warr Pavilion, 1935

ST / SG + 2G (or IGS)
Frame repair + single-pane glass + secondary glazing

Van Nelle Factory, 1931

Zeche Zollverein, 1932/1961

Viipuri Library, 1935

Hardenberg House, 1956

REPLACEMENT INTERVENTION TYPES

AL / SG
Non-thermally broken aluminum frames + single-pane glass

Bauhaus Dessau, 1926

Convent of La Tourette, 1960

ST / SG
Non-thermally broken steel frames + single-pane glass

Sanatorium Zonnestraal, 1928–1931

Lever House, 1952

S. R. Crown Hall, 1956

ST / IGU
Non-thermally broken steel frames + IGUs

Fagus Factory, 1912/1925

Sanatorium Zonnestraal, 1928–1931

Zeche Zollverein, 1932/1961

City of Refuge, 1933/1953

ST-TB / IGU
Thermally broken steel frames + IGUs

Bauhaus Dessau, 1926

Zeche Zollverein, 1932/1961

Solomon R. Guggenheim Museum, 1959

AL-TB / IGU
Thermally broken aluminum frames + IGUs

Yale University Art Gallery, 1953

Hardenberg House, 1956

Villa Tugendhat

Brno, Czech Republic
Ludwig Mies van der Rohe, 1930

Rear (southwest) facade of the Villa Tugendhat in Brno, Czech Republic, view from garden, 1930.

Garden stoop and dining room window wall at the rear facade (note the large size of the single-pane glazing), 1932.

The Villa Tugendhat in Brno's residential neighborhood of Černá Pole is a three-story residence designed by German-American architect Ludwig Mies van der Rohe (1886–1969). Located at the top of a sloped site with sweeping vistas, it was commissioned by Grete (1903–1970) and Fritz Tugendhat (1895–1958) in 1928 as their family home. Completed in 1930, the building is the result of the couple's longing "for a modern spacious house with clear and simple shapes."[1] Mies' design solution, based on a steel frame construction (unusual for a home at that time), afforded many advantages such as thin non-load-bearing walls, an open plan with a different configuration on each floor and large glazed walls that opened the rooms to the exterior. The top floor of the house, which includes the bedrooms and the partially covered upper terrace, is accessed from the street level. From there, a staircase enclosed within a floor-to-ceiling steel frame window wall with milky, frosted glass leads down to the intermediate level below where the living room, dining room and kitchen are located. A secondary spiral stair leads from the kitchen down to the service and mechanical rooms in the basement. The dining room leads to another covered terrace which has a wide stoop leading down to the garden and on to the landscaped slope below.

Replacement window wall (installed after WWII) with significant damage from leakage, condensation and frost, 1981.

Along with the extensive use of high-quality materials such as white travertine floors, a wall of honey-colored onyx from Morocco and vivid veneers made of exotic woods from south-east Asia, the extensive use of metals is one of the most character-defining features of the house. In addition to the slender cross-section steel columns made of riveted angles, the house includes brass cladding with a bronze patina coating at the column covers on the upper terrace, and brass chromium plating with a bright luster at the column covers in the main living area.[2] All of the window frames and exterior doors were made of steel, allowing the framing members to be as thin as possible. The way that the steel frame glazing connects the interiors to the surroundings is also a unique feature of the Villa. At the main living space on the second floor, the southeast and garden facades are glazed from floor to ceiling. Two large 16'–5" x 9'–10" (5m x 3m) retractable steel frame windows with a counterweight system operated by an electric motor opened the space into the garden and landscaped slope below (chrome-plated toe guards and a guardrail provide for fall protection and retractable fabric awnings provided sun protection). The design of the house featured other technical innovations such as a forced-air ventilation system serviced by ductwork located at the basement level below, which provided a combination of heating, air-cooling and humidifying. The house also included a chamber built expressly to store furs and other clothing.[3]

CONDITIONS PRIOR TO INTERVENTION

The building was used as a residence by the Tugendhats from 1930–1938. The family left Brno in 1938, and at the beginning of October 1939 the Villa was taken over by the German Gestapo. During WWII, it was used as residence and office of the director of

View through main living space into the overgrown winter garden before restoration, 2010.

the Klöcknerwerke, a German steel manufacturer with operations in Brno. These changes in stewardship and occupancy led to radical alterations. At the street facade, the frosted glass window wall and the upper terrace entryway were bricked up. Additional walls separating the interior were inserted. In April 1945 the Soviet Army contributed to the devastation of the building, using the interior for quarters and stables, burning the wood furniture and destroying the floors. Subjected to blast loads during bombardment, all the exterior glazed walls were broken, except for the retractable window opposite to the onyx wall.[4]

Following the end of the war in Europe, modern dancer Karla Hladká arranged with Brno architect Albín Hofírek to repair the Villa prior to using it for her own private dance school, which she opened in August 1945.[5] As part of this repair campaign, broken window panes were temporarily glazed and the linoleum floor in the main living area was replaced with cement board. The Villa was used as a private dance school until 1950 when its ownership was transferred to the state of Czechoslovakia. In 1955, a rehabilitation center for children with spinal defects was established at the former Villa, lasting until the late 1960s.[6] By then, the sliding walls damaged during the war had been replaced with fixed multi-lite storefronts. In 1967, Brno architect František Kalivoda visited the Villa and met with Grete Tugendhat with the strong intention of working on its renovation. This renovation attempt was halted by

Rear (southwest) facade with the 1980s-era replacement window wall in poor condition (corroded, warped and inoperable due to lack of maintenance), ca. 2010.

Existing front (northeast) facade window wall before restoration showing two acrylic glass panes per bay joined by silicone putty (originally the glass was frosted single-pane), 2010.

the occupation of Czechoslovakia by the former Soviet Union in 1968 coupled with Mies' passing the following year, Mrs. Tugendhat's death in 1970 and Mr. Kalivoda's a year later.[7]

The Villa became the property of the City of Brno in 1980. The first overall renovation of the building, aimed at converting it into a guest house for the City of Brno, began in 1981 and was completed in 1985. The project, prepared by the State Institute for Reconstruction of Historical Towns and Buildings in Brno, was problematic from the start. The renovation architects only made use of archival plans and photographs in the collections of the Brno City Museum, without consulting other original project sources such as the Mies archives. In addition, no construction history, restoration plans or materials conservation research were undertaken. Original elements were destroyed when construction work was performed prior to receiving the design documents. Disputes over construction changes and other issues with the construction com-

Side (northwest) facade during restoration, 2011.

pany of the City of Brno (the materials supplier and contractor) worked against the success of the project.[8] This was evident on the exterior work that was performed. For example, after the glass from a foreign supplier was rejected, the replacement of the original large single-glass panes was carried out using two glass sheets butt-joined with clear silicone sealant. The large glass pane at the only surviving retractable window was removed as it did not match the replacement. Steel window and door frames showed evidence of significant corrosion due to thermal bridges and condensation, yet they were merely cleaned, repaired and painted.[9]

In 1994, Villa Tugendhat was placed under the authority of the Brno City Museum and for the first time it opened to the public as a house museum.[10] It was only after its designation as a World Heritage Site by UNESCO in 2001 that in-depth Construction History Research (CHR) was undertaken.[11] Under the supervision of Karel Ksander, the CHR was completed in 2005. Concurrent restoration research was conducted from 2003–2005 under the supervision of conservation specialist Ivo Hammer. Both studies served as a basis for the more recent (and careful) restoration of the Villa that was carried out 2010–2012. In 2012 the Villa became part of the international Iconic Houses network and was reopened to the public as a house museum.[12]

Rear (southwest) facade during restoration, 2011.

INTERVENTION

The 2010–2012 restoration of the Villa aimed to return the building to its original historic state as designed and built in 1929–1930. This meant maintaining the authentic material composition of the building through, firstly, rescuing and prolonging the life of the monument as "a preserved original", and, secondly, preparation and reconstruction of the house in its original form.[13] An international expert committee of specialists, the Tugendhat House International

Handling of replacement frosted glass during restoration, 2011.

Committee (THICOM), was assembled in 2009 to advise the City of Brno with regard to the evaluation of technical issues concerning "the preparation and implementation of the Tugendhat House according to the principles of the preservation of a monument."[14]

The materiality and aesthetic of the original steel window frames was described by Ivo Hammer in 2008 as follows:

> The exterior sides of the metals, e.g. the frames of the windows, were originally coated not only with a bluish grey oil based paint on top of several preparatory layers, but also with a clear varnish (possibly acetate of cellulose). This type of varnish application is unusual and is not necessary for protection, but has only an aesthetic intention. It gives the metal color a greater saturation and in [sic] the same time somewhat suggests a metallic surface. Certainly, it is not unintentional that the tonal value of this paint is similar to the tone of the oxidized lead which is protecting the bases of the window frames.[15]

Installation of replacement bent single-pane frosted glass at front facade, 2011.

The glazing and the paintwork of the steel elements were challenging elements in the restoration. Both of the retractable windows at the garden facade were considerably warped, and the one opposite the dining room was inoperable. Fortunately, the original mechanisms for opening the windows were fully preserved and were overhauled and restored to full working order.[16] Restorers removed the secondary paintwork from these structures as carefully as possible, leaving any preserved layers of original paint untouched. In accordance with the restoration plan, the paint and coating work was divided into three basic categories: (1) completely deteriorated paintwork, (2) partially deteriorated paintwork, and (3) preserved paintwork.

Fragments of original glass from the window walls were found and analyzed, and the raw material remanufactured and further processed to produce an authentic replica. This involved cutting the glass to the required size and shape of the panes (many of them were bent) and finishing the edges. The handling of the large-format glass panes during manufacturing, transportation, site delivery and installation was one of the most demanding tasks in the Villa's restoration.[17]

Interior view of restored window wall with frosted glass at main entrance, 2012.

1 Fixed steel frame window wall
2 Retractable steel frame window wall
3 Motorized mechanism for retractable
   window wall
4 Single-pane clear glass
5 Retractable awning
6 Single-pane frosted glass

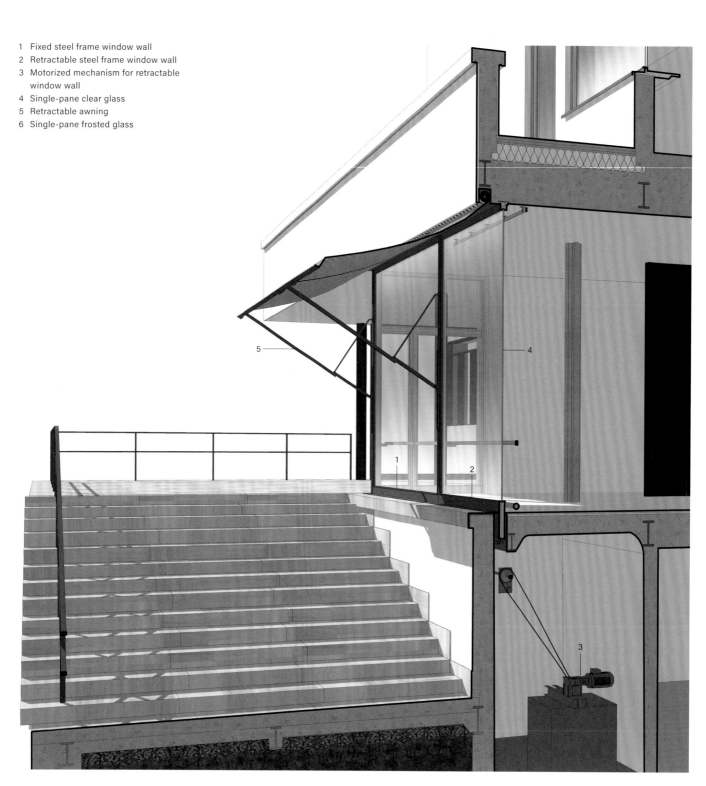

Section at rear (southwest) facade steel frame retractable window wall with fabric awning.

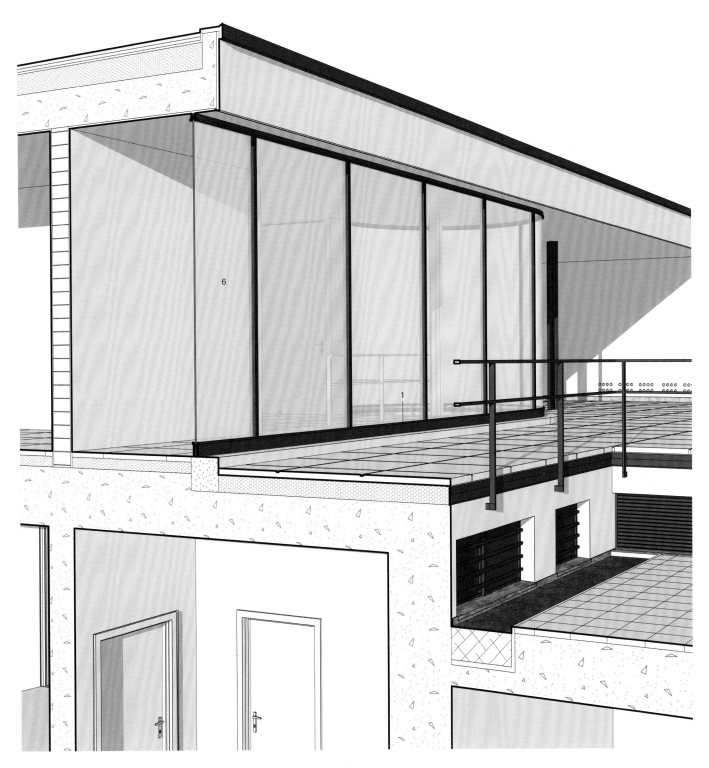

Section at front (northeast) facade steel frame window wall with frosted glass.

Garden view of rear (southwest) facade after restoration, 2012.

Rear (southwest) facade, view of window wall looking into the main living space after restoration, 2012.

Restored original steel frames and door hardware, 2012.

Interior view after restoration of the steel frame window walls and ebony wall, 2012.

The original frosted glazing at the entrance hall on the street facade included sandblasted glass with a smooth and glossy exterior and a matte interior. For technical reasons this was substituted with satin glass, which is not sandblasted on the inside but chemically treated.

COMMENTS

The decision to establish the expert committee THICOM to advise the City of Brno was an important element of the restoration process—and the key to avoiding the inappropriateness of the earlier interventions. The work of the committee was based on the careful construction history research and the restoration research that is still available on site and online at the Study and Documentation Centre (SDC-VT). The restoration decisions were driven by the fact that the Villa is used as a house museum, with specific authenticity requirements, and also subject to a large number of visitors. This led to a very careful restoration of the retractable mechanism and to a sensitive reproduction of the original glass qualities based on chemical analysis of surviving samples. With regard to the building's materials and surface treatments, Dagmar Černouškova from the Brno City Museum SDC-VT praised the sensitive and accurate renewal of all original surfaces and elements during the restoration, writing:

> The basis of the villa restoration was the complete restoration of the building, [with] as sensitive and accurate renewal of all original surfaces and elements as possible. In correspondence with this, the manner of replenishment of the missing house contents such as furnishings, built-in and other furniture was set... The renovators took special care of the surfaces. The original substances of the individual material elements were preserved as good as possible, especially the outer and inner plasters, metallic surfaces and wood.[18]

Front (northeast) facade after restoration, 2012.

Underlying this work, the extensive amount of detailed research and documentation performed serves as a high standard worth striving for on any restoration project. The work also benefitted greatly from the extended network of experts involved who contributed to the development of very particular and refined solutions. The repair and restoration of the steel-framed window elements and their retractable mechanisms, for instance, demanded substantial craftsmanship, and the outcome is outstanding.

Overall, the restoration of Villa Tugendhat was a careful intervention that reinstated the original appearance and functionality of its architecture, interior design and furnishings, reviving the original character and atmosphere present during the time it was used as the Tugendhat family home. The restoration has been

CREDITS

Villa Tugendhat (1930)
Brno, Czech Republic

ORIGINAL CONSTRUCTION

Architect
Ludwig Mies van der Rohe

Contractors
Artur and Mořic Eisler

1981–1985 INTERVENTIONS

Architect
The State Institute for
Reconstruction of Historical
Towns and Buildings in Brno
prepared the project

Design Team
Kamil Fuchs (team leader),
Jarmila Kutějová, Josef Janeček,
Adéla Jeřábková

2005–2012 INTERVENTIONS

Collaborating Architects
Omnia Project, Archteam and
RAW

Design Team
Marek Tichý (specialized guarantor of the project, authorized expert in restoration of buildings), Milan Rak, Tomáš Rusín, Petr Řehořka, Alexandr

Skalický, Ivan Wahla, Vítek
Tichý (main engineer and
project coordinator)

**Glass Manufacturer, Glass
Processing and Glazing**
Saint-Gobain Glass Benelux
(Belgium), Ertl Glass AG
(Austria), Isosklo, spol. s r.o.
(Czech Rep.), DIPRO okna,
s. r. o. in collaboration with
Uplifter, s. r. o. (Czech Rep.)

View at dusk of rear and winter garden facades after restoration, 2012.

successful not only in terms of conservation, but also as it relates to cultural history. The restored villa is a showcase of experimental concepts in living, technology and comfort; one whose habitability was critically discussed shortly after its construction. It also conveys the commitment of the architectural team and THICOM, who together with the heirs of the original inhabitants, the museum and the City of Brno worked together to ensure a successful project completion.[19]

NOTES

1 Villa Tugendhat, "The Commissioners," accessed 16 February 2019, http://www. tugendhat.eu/en/villa-tugendhat/the-commissioners-.html.

2 Villa Tugendhat, "The Materials," accessed 16 February 2019, http://www.tugendhat.eu/en/the-building/the-materials.html.

3 Iconic Houses, "Technical Innovations," accessed 10 July 2018, https://www.iconichouses.org/specials/villa-tugendhat/technical-innovations.

4 Villa Tugendhat, "War Devastation," accessed 16 February 2019, http://www.tugendhat.eu/en/war-devastation.html.

5 Villa Tugendhat, "The Building – After the Departure of the Family," accessed 10 July 2018, http://www.tugendhat.eu/en/after-the-departure-of-the-family.html.

6 Villa Tugendhat, "Physiotherapy Centre," accessed 17 February 2019, http://www.tugendhat.eu/en/physiotherapy-centre.html

7 Ludwig Mies van der Rohe died in 1969, Grete Tugendhat in 1970 and František Kalivoda in 1971.

8 Villa Tugendhat, "The Building – The Restoration of the Villa from 1981 to 1985," accessed 10 July 2018, http://www.tugendhat.eu/en/the-building/the-restoration-of-the-villa-from-1981-to-1985.html.

9 Ibid.

10 Iveta Černá, Dagmar Černoušková, "Mies in Brno: the Tugendhat House," Brno: Brno City Museum, 2018.

11 The Construction History Research (CHR) was completed under the supervision of Karel Ksander, with a team of specialists who were already responsible for the CHR of the Müller Villa (Adolf Loos) in Prague; Villa Tugendhat, "Research and Publications," accessed 10 July 2018, http://www.tugendhat.eu/en/research-and-publications/construction-history-research.html.

12 The network Iconic Houses (IH) was established in 2012 by Natascha Drabbe (in cooperation with Iveta Černá, Kimberli Meyer and Lynda Waggoner). For more information on IH, see www.iconichouses.org.

13 Iconic Houses, "Restoration," accessed 10 July 2018, https://www.iconichouses.org/specials/villa-tugendhat/restoration.

14 Villa Tugendhat, "THICOM," accessed 10 July 2018, http://www.tugendhat.eu/en/science-and-research/thicom.html.

15 Ivo Hammer, "The Original Intention – Intention of the Original? Remarks on the Importance of Materiality Regarding the Preservation of the Tugendhat House and Other Buildings of Modernism," accessed 13 April 2008, http://www.tugendhat.eu/data/Ivo_Hammer_Original_%20intention_2008.pdf.

16 Based on information provided in 2016 by Dr. Iveta Czerna, Director of the Villa Tugendhat Museum.

17 Replacement glass was produced at the Auvelais plant of the Saint-Gobain Glass Benelux glassworks, finishing was performed by Ertl Glass AG and Isosklo, spol. s.r.o. and the glazing installation itself was carried out by DIPRO okna, s.r.o., in collaboration with Uplifter CZ, s.r.o.

18 Dagmar Černoušková, Alena Štěpánková, Marek Navrátil, "Villa Tugendhat is once again close to the original," *BRNO BUSINESS & STYLE* 20, no. 1 (2012): pp. 4–7, 86, accessed 10 July 2018, http://www.tugendhat.eu/data/BB-EN.Tugendhat.pdf.

19 Daniela Hammer-Tugendhat, Ivo Hammer, Wolf Tegethoff, *Haus Tugendhat. Ludwig Mies van der Rohe* (Birkhäuser: Basel, 2014 (new edition), pp. 29–34.

Glass House

New Canaan, Connecticut, USA
Phillip Johnson, 1949

Interior view of the Glass House during construction in New Canaan, Connecticut, ca. 1949.

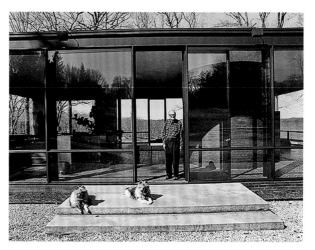

Front (east) facade with Phillip Johnson standing in the main entrance, ca. 1998.

The Glass House and its lesser-known counterpart, the brick Guest House, were designed between 1945 and 1947 by American architect Philip Johnson (1906–2005) as his weekend residence, and completed two years later. Situated away from the street and along the edge of a cliff, the Glass House overlooks a pond set in a bucolic landscape. Its design was inspired by Mies van der Rohe's 1945 design for the Farnsworth House in Plano, Illinois (see page 82), which was completed in 1951 and had been included in an exhibition about Mies' work at the Museum of Modern Art organized by Johnson.[1] In contrast to the Farnsworth House, which is elevated on steel stilts, the Glass House is set on the ground. It is also more modest in size and scale, and its interior was designed entirely by the architect.

The house consists of a steel frame structure with exterior columns at the corners and window walls with large glass panes separated by steel mullions and transoms. The primary structure of the building and window walls is constructed of wide-flange steel columns and concealed H-beams with exposed steel channels. The 10'-5" (3.18m) high steel frame window wall is single-glazed with 1/4" (6mm) thick plate glass panes, the largest of which are 12'-9 3/4" (3.91m) to 17'-10 1/2" (5.45m) wide and 7'-10 1/8" (2.39m) tall.[2] Mullions and transom bars, as well as the smaller glass stops, are made of 1" (2.5cm) wide by 3 3/4" (9.5cm) deep rectangular steel stock bars. All of the steelwork is finished with a black paint coating.

Front (east) facade, 2018.

Unlike the Farnsworth House, which was located on a remote Midwest plain and accessible only to those few visitors invited by its reclusive owner, the Glass House immediately became a social hub, suiting its extrovert owner and architect. Johnson's integration of design elements at the Glass House, where enclosure, views, furniture and space are unified cohesively, earned the House and its glazed steel frame a reputation as one of the most influential Modern designs of US post-war architecture. It also unleashed an enduring fascination about a Modern architect's own house becoming his autobiography.[3]

CONDITION PRIOR TO INTERVENTION

Over the years, some of the original glass panes broke and were replaced under Johnson's supervision and later under that of his designated caretaker, who remained at the property for some time after both Johnson and his partner, David Whitney, passed away.[4] Interestingly, the recurring glass breakage was attributed to local wild turkeys crashing into the house after spotting their reflection and rushing at the windows in a defiant act of territorial protection, or perhaps because they simply did not see the glass.[5] During this period, broken panels of 1/4" (6mm) thick plate glass were replaced with new 3/8" (9.5mm) float glass.[6] Although slightly thicker, due to the manufacturing standards for the large sizes required for the Glass House, float glass had became more readily available and cost effective than the original plate glass. It also

TIMELINE

1945	Schematic design begins
1946	Philip Johnson purchases land in New Canaan
1947	Philip Johnson finalizes design
1949	Construction of the Glass House completed
1986	Philip Johnson bequeaths the Glass House to the National Trust for Historic Preservation and retains a life estate
1997	Listed on the National Register of Historic Places and designated a National Historic Landmark
2005	National Trust for Historic Preservation becomes the steward of the Philip Johnson Glass House
2007	National Trust for Historic Preservation opens the Glass House to the public

Interior view looking west out of the rear facade towards the pond, 2006.

afforded a flatter surface with less visual distortion, a change that Johnson seemed to have embraced alongside the many additions and alterations he designed for the Glass House site over the decades.

INTERVENTION

Since the property was placed under the care of the National Trust for Historic Preservation in 2005, at least five glass panes have been replaced using 3/8" (9.5mm) tempered glass, which has become the standard replacement glass specification for the Glass House.[7] Other than these localized glass replacement interventions, the House and other buildings and artifacts on the property are maintained as a house museum, and architectural features, furniture and artifacts are preserved in their original condition.[8] According to the National Trust, the maintenance plan for the house includes cleaning on a weekly basis and, as needed, alternate window washing (interior one week, exterior the next week), painting approximately every five years, removal and replacement of all sealants on the exterior, localized sealant removal and replacement where needed on the interior and surface preparation and painting as needed. The house is, however, in need of rust removal and steel frame repairs, and replacement of door closers and saddles, among other pressing interior restoration needs.[9]

Southwest corner of the facade, 2006.

COMMENTS

The shift from the original 1/4″ (6mm) plate glass to the thicker 3/8″ (9.5mm) float glass, and then later from 3/8″ (9.5mm) float glass to 3/8″ (9.5mm) tempered glass is the result of evolving industry standards for glass manufacturing. The current use of tempered glass is also a response to safety concerns now that the site is a public venue and contemporary building codes mandate the use of safety glass for installations within 18″ (46cm) of the finished floor. Even when the location of original and replacement glass is well documented, the change to tempered glass comes with its own challenges related to the visual distortions of the roller-wave effect that characterizes tempered glass.

The large glass panes are one of the most defining features of the Glass House. In designing the Glass House and surrounding structures on the property, Johnson was interested in exploring a sense of vulnerability and fragility in relation to place. Reflecting in 1993 on how he strived to elicit feelings of "nostalgia, precariousness and sexual yearning" in his work and to transfer those feelings into stone and steel, writing that, at the Glass House, one feels like "the glass might shatter in the wind!"[10] The large panes convey Johnson's interest in challenging the traditional role of architecture as a provider of shelter, security and comfort. Glass makes up roughly 90% of the exterior vertical enclosure,

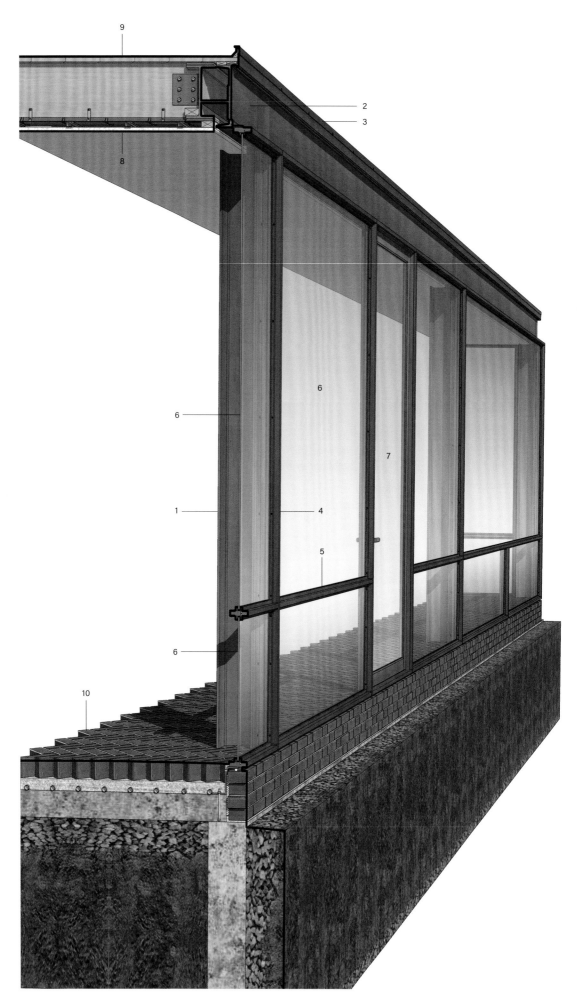

Typical section at existing steel frame window wall.

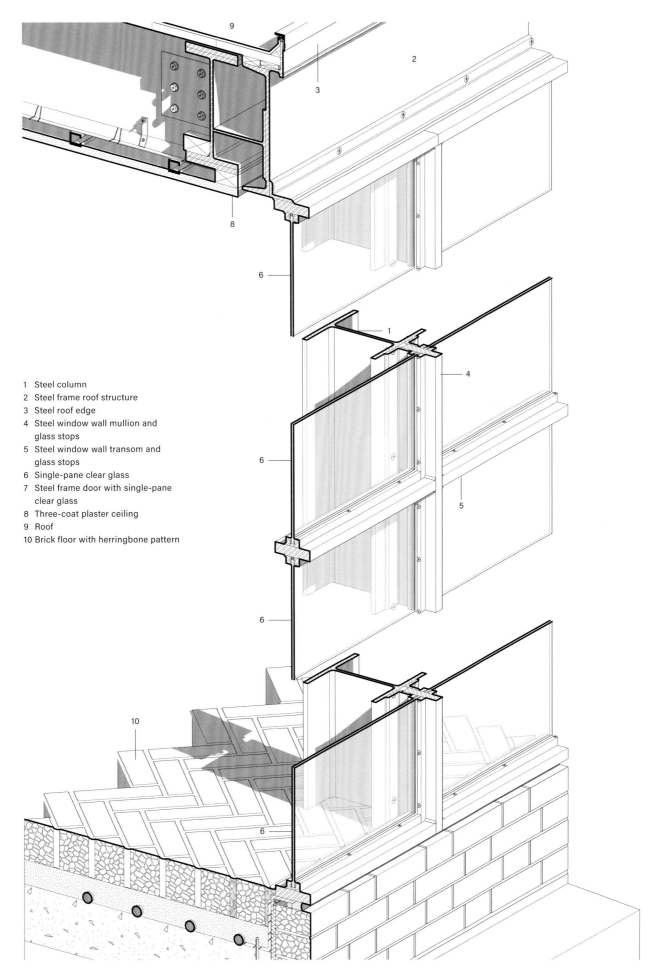

1 Steel column
2 Steel frame roof structure
3 Steel roof edge
4 Steel window wall mullion and
  glass stops
5 Steel window wall transom and
  glass stops
6 Single-pane clear glass
7 Steel frame door with single-pane
  clear glass
8 Three-coat plaster ceiling
9 Roof
10 Brick floor with herringbone pattern

Typical existing window wall details.

Steel being prepared for painting, undated image.

which makes the properties of the glass (its color, reflectivity, transparency and level of visual distortion, for example) significant character-defining features. David Whitney and Jeffrey Kipnis encapsulate this using Mies' own thoughts:

> The multiple reflections on the 18' pieces of plate glass, which seem superimposed on the view through the house, help give the glass a type of solidarity; a direct Miesian aim which he expressed twenty-five years ago: "I discovered by working with actual glass models that the important thing is the play of reflections and not the effect of light and shadow as in ordinary buildings."[11]

This seems to explain why Johnson embraced the shift from the original plate glass to float glass in his own time—and also supports the appropriateness of the latest shift from float to tempered

CREDITS

Glass House (1949)
New Canaan, Connecticut

ORIGINAL CONSTRUCTION

Architect
Philip Johnson

1949-2005 INTERVENTIONS

Architect
Philip Johnson

2005-PRESENT INTERVENTIONS

Property Caretaker (2005)
National Trust for Historic
Preservation (2005–Present)

Owner
National Trust for Historic
Preservation

Front (east) facade, view looking south, 2008.

glass. For the Glass House, change is part of the conceptual and physical history and evolution of the building and the site. What is most valuable to preserve at the Glass House is therefore not only the glass itself, but how it reflects and filters light and how it characterizes the building's relationship to the site, to its former architect-resident and his partner, and to all the social, design and architectural events that it witnessed for decades. This is a case where maintenance is the most appropriate reglazing intervention. The curatorial plan set forth by the National Trust has succeeded with this for the last decade and promises to continue to bring to the Glass House a well-balanced mix of conservation-minded interventions and cultural programming.

NOTES

1 Alice T. Friedman, *Women and the Making of the Modern House* (New Haven, CT, Yale University Press, 2006), p.130, accessed 23 December 2018, https://books.google.com/books/yup?id=-WXuEAwKcSQC&lpg=-PP1&pg=PA130#v=onepage&q&f=false.

2 National Trust for Historic Preservation, "The Glass House Architectural Drawings," accessed 5 June 2016, http://theglasshouse.org/learn/architecturaldrawings/.

3 National Trust for Historic Preservation, "Architecture & Influence – The Phillip Johnson Glass House Oral History Project," online video, 3:02 mins, accessed 18 August 2018, http://theglasshouse.org/explore/the-glass-house/video/.

4 Fred A. Bernstein, "Treading Gently on Hallowed Ground," *The New York Times*, 31 August 2006, Habitat Section.

5 "Coyotes not talking turkey at Johnson's Glass House," *Chicago Tribune*, The Advocate, 10 December 2006.

6 Gregory Sages, email message to author, 2 August 2016.

7 Ibid.

8 A recent exception is the replacement of the original three-coat plaster on metal lathe ceiling, which was abated (it contained asbestos) and replaced in-kind to address decay and reinstate door operability, which had been compromised. For further information see: The Glass House, "Replacement of the Glass House Ceiling," accessed 6 April 2019, http://theglasshouse.org/learn/replacement-of-the-glass-house-ceiling/.

9 Sages, email to author, 2 August 2016.

10 David Whitney and Jeffrey Kipnis, eds., *Phillip Johnson: The Glass House* (New York, Pantheon Books, 1993), p.ix.

11 Ibid., p.13.

Hallidie Building

San Francisco, California, USA
Willis Polk, 1918

Hallidie Building in San Francisco, California, view of the main facade on Sutter Street, ca. 1970.

The Hallidie Building in San Francisco's Financial District was designed by Willis Jefferson Polk (1867–1924), a US architect who for ten years was the West Coast representative of D. H. Burnham & Company. The building was commissioned by the University of California, and named for Andrew Smith Hallidie, a university regent and engineer born in England who is credited with inventing the local cable car.[1] The seven-story building has the distinction of having the first application of an all-glass curtain wall (with steel frames, but no metal or masonry spandrels) of any building in the US. The revolutionary steel frame curtain wall at the front (south) facade on Sutter Street was built using a combination of steel angles, T-sections and plates, as well as iron railings and steel brackets combined with neo-Gothic ornaments made of cold-pressed sheet metal with a painted gold finish.[2] A precedent to Polk's design for the front facade of the Hallidie Building can be found in the Boley Building in Kansas City (Louis S. Curtiss, 1909), where Polk had lived and was reportedly a member of the Kansas City Architects Sketch Club alongside Curtiss.[3] While the Boley Build-

56

Deterioration of second-floor curtain wall frame (left side of image) and decorative sheet metal (right side of image), ca. 2010.

Typical condition of the deteriorated curtain wall windows prior to restoration, ca. 2010.

ing window wall is comprised of cast iron mullions and spandrel panels supported by a riveted steel channel that covers the edge of the cantilevered reinforced-concrete slab, the Hallidie Building curtain wall is supported by steel outriggers attached to a thin cantilevered slab, which Polk's drawings refer to as a "window ledge." The curtain wall at the Hallidie Building includes horizontally pivoting steel frame windows with 1/4" (6mm) thick plate glass at each floor whereas the Boley Building includes operable units at only a few window bays.

At the Hallidie Building, the alignment of the curtain wall glass surfaces with the facades of the adjacent masonry buildings (rather than being set back a few inches within masonry surrounds) was prized since its inception for allowing light to penetrate deep into the building, thus maximizing the usable areas of the interior spaces. After the building opened, its unique separation of structure and glazed steel frame facade was received with mixed results, as evident in this 1918 excerpt from *American Architect:*

> The architectural treatment of this type of building presents a new problem to the architect. In this case the architect has designed the exterior envelope without attempting to preserve a concordance with the structural skeleton, except in some horizontal members. This divorcement of the external covering from the skeleton is very unusual and by its adoption the designing of the elevation is done with a freedom not otherwise possible.[4]

The divorce of the curtain wall from the building structure, along with its translucency and light weight were unprecedented (in the US at least). This departure from the traditional design methods that had defined the architectural design vocabulary for centuries is precisely where the building's significance stems from. Behind the decorative sheet metal ornaments and ferrous balconies and fire escapes, the steel frame curtain wall at the Hallidie Building delineated the facade typology that was to dominate Modern architecture in the US and beyond during the 20th century.

TIMELINE

1918	Built
1918– 2011	Localized window replacement
1971	Listed on the National Register of Historic Places, the California Register of Historic Resources, and designated as San Francisco City Landmark Number 37
2010	Existing conditions assessment and repair recommendations by McGinnis Chen Associates, Inc. and Page & Turnbull, Inc.
2011– 2012	Phase 1 restoration work (restoration of the decorative frieze panels, sheet metal details, metal railings, structural framework, fire escape ladders and structural steel I-beams)
2013– 2014	Phase II of restoration work (structural work, curtain wall repairs, window repair and replacement and implementation of weather protection measures)

Looking up at the curtain wall after restoration, ca. 2013.

CONDITION PRIOR TO INTERVENTION

Over the years, the steel frame curtain wall warped due to a lack of thermal expansion joints, among other factors. The steel outriggers and the I-beams supporting the balconies had all corroded. The corrosion at the connection of the ferrous balcony railings to the steel window mullion system was extensive and the original steel brackets had deteriorated to the point where they were no longer functional.[5] The original steel outriggers, designed to bear the vertical (gravity) loads of the curtain wall, were located at every other mullion, leaving the remaining mullions unsupported.[6] This resulted in 3/8" (9.5mm) to 3/4" (19mm) of settling in relation to the supported mullions. It was also discovered that the entire curtain wall had no lateral (wind) support system.[7] By 2013, a multidisciplinary team led by architects McGinnis Chen Associates, Inc. and preservation consultants Page & Turnbull, Inc. indicated that it was "just a matter of time before portions of the facade supported by these brackets will fall off the building."[8] Amid the growing concern over the structural condition of the aging ferrous curtain wall the same team was charged with preparing an overdue renovation of the facade.

INTERVENTION

A phased restoration plan was devised by the McGinnis Chen and Page & Turnbull team. The first phase, approved by the local Historic Preservation Commission in 2011, included the restoration of the decorative frieze panels, sheet metal details, metal railings,

Severely corroded curtain wall window frame (note the deformation caused by rust jacking) prior to restoration, ca. 2010.

Detachment of the vertical steel plate from the corroded window frame prior to restoration, 2010.

structural framework, fire escape ladders and structural steel I-beams. Additional work including the rehabilitation of a limited number of existing steel windows at the curtain wall assembly was approved in 2012. The second phase, approved in 2013, included a combination of structural work, curtain wall repairs, window repair and replacement and implementation of weather protection measures. Under the structural scope, inconspicuous supplemental fin-shaped structural supports were installed every other floor at the unsupported mullions. New steel outriggers were then installed at the unsupported mullions of the opposite floors in order to provide vertical and lateral support for the curtain wall. Matching bolts with a higher structural capacity replaced the original existing ones. At the curtain wall, the warped cover plates at the mullions were removed and replaced with new ones to match the originals, with the added inclusion of a splice joint to allow for expansion and contraction. The window work included repair and in-kind replacement of the window steel frames and sash where required and replacement of the existing 1/4" (6mm) plate glass with single-pane laminated safety glazing. The scope included welding at the corners of the existing window frames and sash, and epoxy installation at the joints of the replacement window frames and sash for additional strengthening and weather protection. Silicone sealants were installed around the curtain wall to prevent water infiltration and new flashing was installed at the windows below the balconies and seventh-floor soffit.[9]

In addition, the steelwork was repainted and the sheet metal ornaments were refinished to reinstate the original golden appearance. All the curtain wall components were primed with Series 90-97 Tneme-Zinc (an aromatic urethane zinc-rich primer) and finished with the original blue color using Series 1071V Fluoronar by Tnemec (a low-VOC semi-gloss fluoropolymer coating) or Series 1072V Fluoronar (also by Tnemec), which is satin finish. The decorative sheet metal was coated with Series 135 Chembuild (a modified polyamidoamine epoxy) and finished with Series 1075 Endura-Shield II (an aliphatic acrylic polyurethane coating), also by Tnemec. The components with the gold finish received a coat

View showing the curtain wall reinforcement installed during the restoration, ca. 2013.

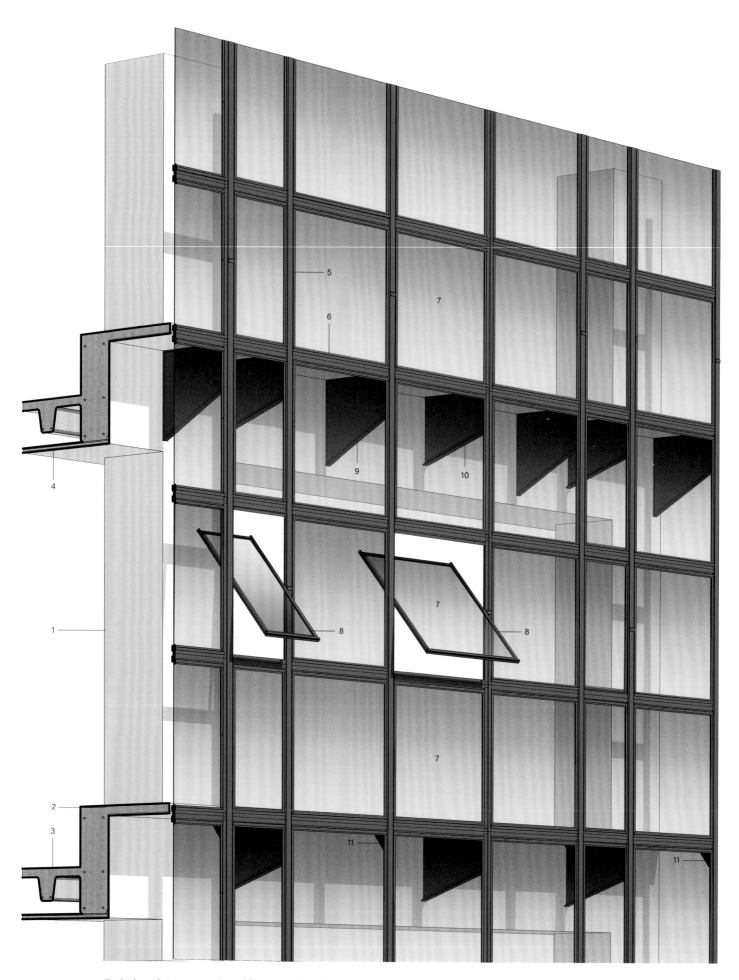

Typical section at restored steel frame curtain wall.

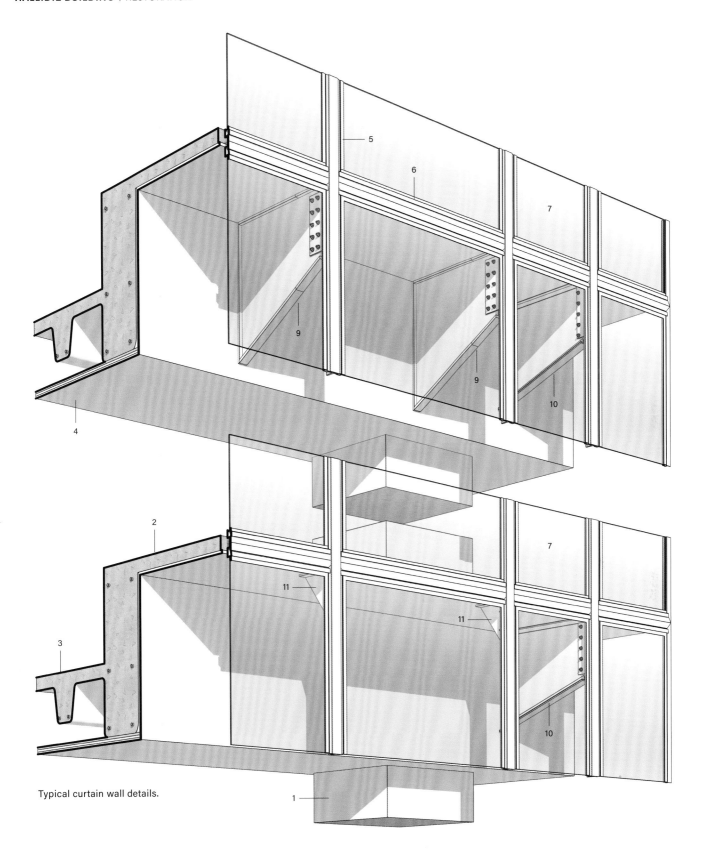

Typical curtain wall details.

1   Reinforced-concrete column
2   Reinforced-concrete girder and window ledge
3   Reinforced-concrete floor
4   Suspended ceiling
5   Steel curtain wall mullion
6   Steel curtain wall transom
7   Single-pane clear glass
8   Steel frame horizontal pivot window
9   Existing steel outriggers
10 New steel outriggers
11 New steel mullion support and wind bracing

Main facade on Sutter Street after restoration, ca. 2013.

CREDITS

Hallidie Building (1918)
San Francisco, California

ORIGINAL CONSTRUCTION

Architects
Willis Polk & Co.

2010–2014 INTERVENTIONS

Owners
Edward J. Conner
Herbert McLaughlin Jr.

Architects
McGinnis Chen Associates, Inc.

Historic Preservation
Page & Turnbull, Inc.

**Structural Engineers
(Curtain Wall)**
Toft, De Nevers & Lee

**Structural Engineers
(Balconies)**
Murphy Burr Curry, Inc.

General Contractor
Cannon Constructors North,
Inc.

Construction
The Albert Group, LLC

Sheet Metal Subcontractor
Van Mulder Sheet Metal, Inc.

Steel Subcontractor
MAS Metals, Inc.

Painting Subcontractor
Abrasive Blasting & Coating,
Inc.

Photography
Sherman Takata

of Series 1078 Fluoronar Metallic, followed by a coat of Series 1079 Metallic Clearcoat.[10]

COMMENTS

The restoration work successfully retained the integrity of the building facade by replacing in-kind the vertical cover plates, repairing the windows and replacing the glass, providing new thermal joints and integrating new supports and flashing within the curtain wall. The installation of new steel outriggers matching the existing ones, coupled with the installation of smaller fin-shaped steel brackets each at alternating floors, allowed the stabilization of the curtain wall by providing lateral bracing and support to the mullions that originally had none. These new structural components, which are visible from the exterior, are finished with the same blue color as the original outriggers and the curtain wall frame. The resulting effect is a combination of new and original components that are indistinguishable from each other, particularly at the floors where the original outriggers were replicated. At these locations, the new and existing outriggers seem too close to each other, which in turn makes them look oversized for their current spacing. What used to be an elegant, rhythmic composition has been turned into ordinary and undistinguished repetition. The outcome, although subtle, distracts from the original design. One way to minimize this effect and make a clear distinction between new and original structural components would have been to use a white finish for the new outriggers and smaller fin-shaped steel brackets so that they blended in with the structure behind. This would have prevented any misinterpretation of the new supplemental structural components as originals.

As the building is listed in the National and State Register of Historic Places, the curtain wall restoration was not required to meet California Energy Code (one of the most stringent in the US). Had the project not been exempt from meeting this code, it is likely that the work would have required removal and replacement of the original steel frame to allow insulated glass units to be installed. This would have doubled the dead load on the slender original steel frame that supports the large glass panes. Either intervention would have significantly altered the building's appearance, detailing and character; therefore, both were rightly deemed inappropriate and not implemented. Considering the extent of remaining original historic fabric and the way in which the restoration work has brought new life to the building facade, the intervention overall is very appropriate for America's first steel-framed curtain wall.

NOTES

1 "U. of C. Building to Bear Hallidie Name," *San Francisco Chronicle* (1869-Current File); 13 December 1917; ProQuest Historical Newspapers: San Francisco Chronicle, p. 4.

2 Annie Lo and Elisa Hernandez Skaggs, "The Hallidie Building: Rehabilitation and Repair of One of America's First Curtain Walls [Abstract CS07b]," paper presented at the Association for Preservation Technology "Preserving the Metropolis" Conference, New York, New York, 11–13 October 2013.

3 Keith Eggener, "Louis Curtiss and the Politics of Architectural Reputation," *Places Journal,* February 2012, accessed 23 January 2019, https://doi.org/10.22269/120206.

4 "An All-glass Front: Some Daylighting Features of the Hallidie Bldg," *American Architect 113* (March 2018): pp. 393–394.

5 Thad Povey and David G. Murphy to Mr. Bruce Albert from Bruce Albert, Inc., Project No. M210-023, 23 March and 13 October 2010, accessed 10 June 2015, http://sf-planning.org/ftp/files/Commission/HPCPackets/Hallidie_Bldg_120110.pdf, 14/25.

6 Tim Frye to Bruce Albert, *memorandum,* January 7, 2013, San Francisco Planning Department, accessed 10 June 2015, http://commissions.sfplanning.org/hpcpackets/2013.0009A.pdf, pp. 2–3, 61/130.

7 Ibid.

8 Povey and Murphy to Albert, 2010.

9 Frye to Albert, *memorandum*, 2013.

10 Tnemec, "Curtain Wall Rises on Iconic Hallidie Building," *News,* accessed 29 December 2018, http://www.tnemec.com/content/news/curtain-wall-rises-on-iconic-hallidie-building#.XCd2SFxKiM8.

Viipuri Library

Vyborg (Viipuri), Russia
Alvar Aalto, 1935

Aerial view of the Viipuri Library in Viipuri (now Vyborg), Russia before WWII, ca. 1930s.

The main entrance hall and adjacent staircase, 1935.

The Viipuri Library was the result of a public competition that took place in 1927, only ten years after Finland gained independence from Russia in 1917. The winning entry was designed by Finish architect Alvar Aalto (1898–1976) whose initial design underwent a series of modifications prior to construction. Several designs were submitted in the years between 1927 and 1933. Construction started in 1934 and was completed in 1935.[1] The design of the library has been described as "a catalogue of innovative concepts which reflect master architect Alvar Aalto's most mature and influential work."[2] Integral to this was its "three-dimensional spatial development [...] in plan and section, where interior spaces were more open, varied and interrelated. To ensure public access to materials, the lending room and reading room were fully integrated."[3] This central space was naturally lit by 57 circular skylights, an innovative design representing a new approach to making knowledge more accessible.[4] The skylights above the lending and reading rooms were each 5'–11" (1.8m) in diameter and consisted of 5/8" (16mm) thick patterned cast glass with a rough surface finish loose-laid over conical concrete drums. Horizontal lites of opalescent glass covered the skylight openings. This daylight-diffusing system was designed to protect the books from direct sunlight and afford an even, glare-free daylight suitable for reading.

The connection of daylight and spatial composition is an overall theme at Viipuri Library, where the windows, skylights and window walls were designed according to the user's spatial needs.[5] Through other specific designs implemented throughout

Steel frame window wall at staircase before WWII, 1935-1936.

Deteriorated window wall at the main staircase, 1991.

the building Aalto improved visual, acoustic and thermal comfort of the interiors.[6]

Like the skylights and the large ribbon windows of the auditorium, the fully glazed staircase window wall of the adjacent lobby is a unique building feature. There, Aalto used a double-glazed steel frame window wall with wooden glass stops made of teak and oak, which insulate the steel frames, thereby reducing heat loss while providing daylight to the adjacent staircase and lobby.

CONDITIONS PRIOR TO INTERVENTION

The Viipuri Library had suffered serious damage and neglect during WWII and the ensuing years. Soviet architects tried to gain access to Aalto's archives in the 1950s in search of guidance for the required repairs, but the political adversity between the USSR and the West made this impossible. In 1952, the first renovation plans were prepared and the building was largely rebuilt between 1955 and when it reopened in 1961. Concerns about the original Modern architectural elements of the library disappearing altogether lasted for years, until the 1980s when attempts were made by Russia (then Soviet Union) to gain access to the original plans and restore the building according to the original design. Since the 1990s some of the spaces have been renovated with the help of the original drawings and other documents conserved in the Alvar Aalto Foundation in Finland. Before Alvar Aalto's office first visited the library in 1991, the original Aalto drawings and material provided by Sergei Kravchenko, a Russian architect who surveyed the building circa 1991, were studied with particular interest.

After 1992, when a detailed condition survey was compiled by Maija Kairamo and Tapani Mustonen, priority was given to repairing the roofs and the skylight, significant components of Aalto's architecture.[7] A historic building survey was first conducted by Joachim Hansson from Alvar Aalto's office in 1994. The various roofs needed new insulation, waterproofing and drainage systems. The skylights had been replaced with plastic domes in

Front view showing (left to right) the restored staircase window wall, main entrance doors and auditorium window wall with ribbon windows above on the upper floor, 2014.

the 1950s/60s and 1990s. Most of the original steel window frames, the interior steel frame window walls and all other exterior doors (except the door to the storage) remained in the building albeit not in good condition. The only exceptions were the lecture hall windows that were made smaller and the window in the storage that had been enlarged. The main entrance bronze doors had been removed and replaced with steel profiles dating to the Soviet period. All original wooden parts had also been removed.

INTERVENTION

A major restoration project from 1994–2013 was coordinated by the Finnish Committee for the Restoration of Viipuri Library, initiated by Aalto's widow, Elissa, after the fall of the Soviet Union in 1991. From 1994–2010, the restoration was undertaken with financial support from both the Finnish and Russian governments, as well as from Aalto admirers from all over the world. In 2010 the Russian government decided to finance the completion of the restoration. From 2010 to 2013, the Finnish Committee was responsible for design and guidance, and the Russian Federal Government for the construction work. The fully restored Municipal Library of the City of Vyborg (Viipuri) was finally inaugurated in 2013.

The work, carried out mainly by Russian contractors under the supervision of Finnish architects, aimed to prevent further deterioration, renew basic facilities and restore the original architectural building components and interiors. Different approaches to the restoration of the building envelope were chosen, but due to the partial and often limited funding the work was implemented according to priorities established as part of the restoration plan. The original in-swing secondary glazing and steel frames of the

Flat skylight replacement in progress, 2003.

Detail of replacement skylight curb and upper laminated glass pane, 2012.

Upper roof, view of restored skylights with replacement clear single-pane laminated, tempered glass and insulated curbs with copper counter flashing, 2012.

window wall in front of the main staircase were carefully restored. All of the glass was replaced in-kind, and the interior sashes were repaired where necessary.

The skylights above the reading and lending rooms were restored back to their original appearance. To improve safety and energy efficiency, laminated tempered glass was used for the upper pane and a lower pane of clear laminated glass was added. The upper laminated panes are made of two layers of 5/16" (8mm) glass and are installed on a new plywood curb atop the original concrete drums. The new curbs are slightly taller than the originals (adjusted for the new insulated roof height) and the newly created cavity is ventilated to avoid condensation. The lower laminated pane, made with two layers of 5/32" (4mm) glass, was installed at the same elevation as the original glass pane.

The steel frames and sashes of the ribbon windows in the auditorium were restored. The original single-pane glass was removed and replaced with butt-jointed single-pane laminated glass. The windowsills, which due to a limit on the size of the glass available during the Soviet period had been raised 7 7/8" (20cm), were lowered to the original height and the steel profiles were extended accordingly.

In the children's department, the original single-pane glazing was replaced with IGUs. The steel frames at the main entrance doors were restored; only missing or deteriorated parts were replaced. The original and Soviet-period steel and glass interior enclosures of the lending and reading hall doors were restored and wooden and sliced birch veneer interior doors were built according to the original specifications.

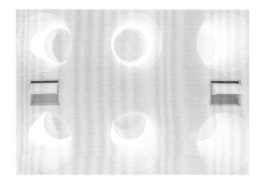

Interior view of skylights after restoration, 2015.

View of the lending and reading rooms, 2015. The skylights supply diffuse daylight that protects the books from direct sunlight and provides an even glare-free light comfortable for reading.

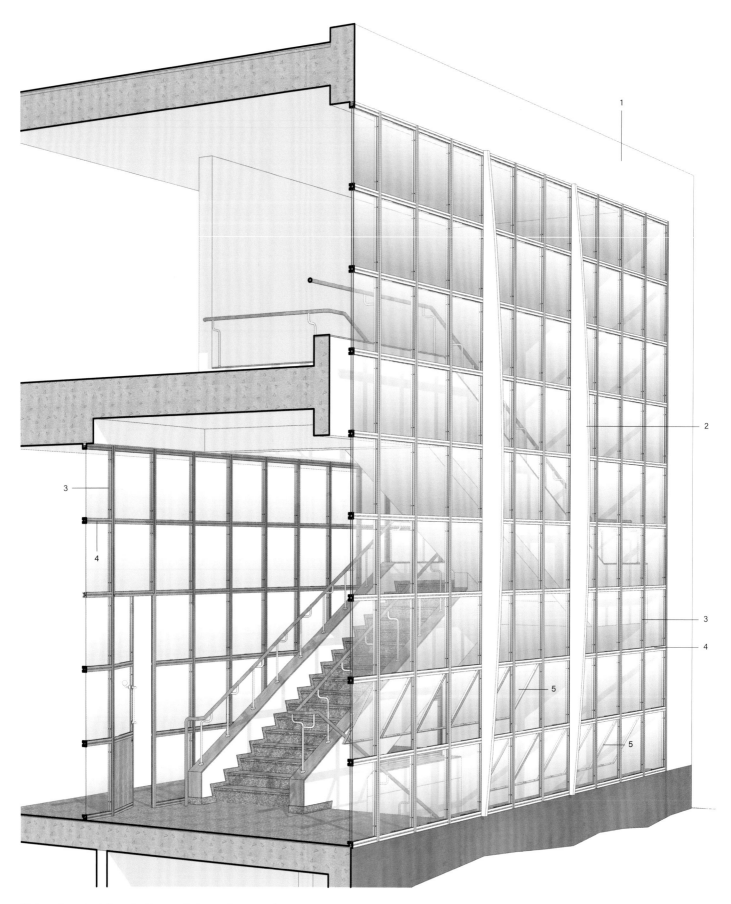

Wall section at existing steel frame window wall.

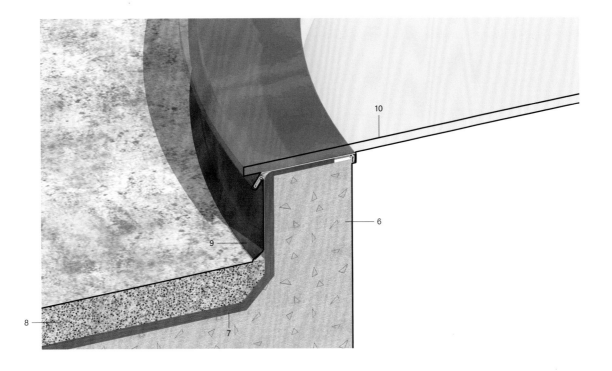

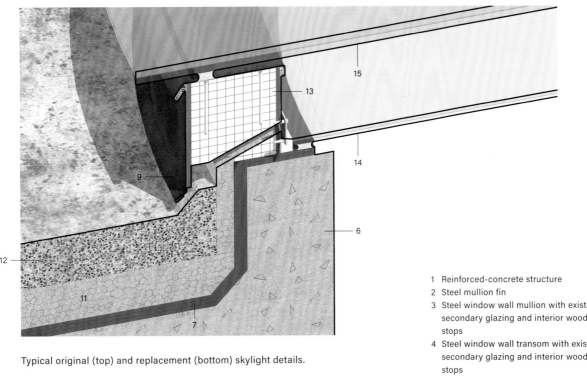

Typical original (top) and replacement (bottom) skylight details.

1 Reinforced-concrete structure
2 Steel mullion fin
3 Steel window wall mullion with existing secondary glazing and interior wood glass stops
4 Steel window wall transom with existing secondary glazing and interior wood glass stops
5 Steel frame in-swing interior awning windows with wood glass stops at existing secondary glazing
6 Reinforced-concrete drums around skylight roof opening
7 Roof membranes
8 Original concrete topping
9 Skylight copper flashing
10 Original single-pane patterned cast glass
11 New roof insulation
12 Concrete topping matching original
13 New insulated and vented skylight curb
14 Replacement single-pane laminated, tempered glass at original patterned glass location
15 New laminated, tempered glass

The thicker bronze frames of the entrance doors contrast with the slim steel frame window wall construction of the main staircase, 2015.

COMMENTS

The history, renovation and restoration of the Viipuri Library is significant for both Modern architecture and 20th-century history. To begin with, Aalto's design went through a transition before becoming an early prototype of a public library. Despite destruction and neglect during and after WWII, the building survived, and went on to be used (although poorly maintained) as a public institution throughout the Soviet era after 1961.[8]

One of the important elements of the restoration process after 1994 was the establishment of an advisory expert committee, the Finnish Committee for the Restoration of Viipuri Library, to advise

CREDITS

Viipuri Library (1935)
Vyborg (Viipuri), Russia

ORIGINAL CONSTRUCTION

Architect
Alvar Aalto

1955–1993 INTERVENTIONS
(Soviet Union and Russian Federation)

Owner
The City of Vyborg, Russia

Architects
Petr Moiseyevich Rozenblum
(from 1950–1957)
Aleksander Mikhailovich Shver
(from 1957–1961)
Sergei Kravchenko
(from 1987–1993)

1991–2013 INTERVENTIONS
(Finnish-Russian Collaboration)

Owner
The City of Vyborg, Russia

Architects
Tapani Mustonen, Leif Englund,
Maija Kairamo

Advisory Board
Finnish Committee for the
Restoration of Viipuri Library
(with Erik Adlercreutz,
chairman and Maija Kairamo,
secretary)

Interior view of steel frame window wall with secondary glazing and wood glazing beads at the Viipuri Library staircase, 2010.

the City of Vyborg (the owner) and the collaborating Russian and Finnish institutions. The work of the committee and the architectural team was based on the careful scientific documentation and construction history research financed by different grants, including the World Monuments Fund, the Getty Foundation and several national funds.

The restoration was able to take advantage of the high-quality original metal constructions that were still in place by collaborating with traditional manufacturers' workshops that were experienced in the restoration of similar projects from that period. Some of the solutions might never have been developed were it not for the skilled independent construction companies willing to take on the liability arising from unconventional and decidedly "not-state-of-the-art" solutions. The specialized restoration architects also used their profound knowledge of the specific materials and technologies (for example, specific alloys that are no longer in use today) in order to make sensible decisions regarding replacement and new material combinations. In the end, Aalto's solutions proved to be adequate and efficient for the purpose they were designed for. Most of them were developed with models in an iterative process before construction started, indicating Aalto used the extra time resulting from the delayed construction start to improve his concepts—something rarely possible today given the time and cost constraints that characterize the contemporary building industry.

The project was honored with the 2014 World Monuments Fund/Knoll Modernism Prize and with the Europa Nostra Awards in 2015. Echoing Aalto, the judges noted that "the building still exists and the architecture has been brought back."[9]

NOTES

1   Laura Berger, "The Building that Disappeared. The Viipuri Library by Alvar Aalto," doctoral dissertation, Aalto University School of Arts, Design and Architecture, 2018, pp. 144, 146; "Alvar Aalto Viipuri Library. 1935," accessed 15 December 2018, http://aalto.vbgcity.ru/node/377.

2   "The Alvar Aalto library in Vyborg," accessed 30 December 2018, http://aalto.vbgcity.ru/node/360.

3   R. Thomas Hille, *The New Public Library: Design Innovation for the Twenty-First Century* (New York: Routledge, 2018), p. 135.

4   Elain Knutt, "Humanity Restored," *Building Design*, 27 January 2006, p. 24.

5   Virginia Cartwright, "Themes of Light: Aalto's Libraries from Viipuri to Mt. Angel," in *Working papers – Alvar Aalto Researchers' Network Seminar 2012*, 12–14 March 2012, Seinäjoki and Jyväskylä, Finland (Helsinki: Alvar Aalto Museum, 2013), accessed 30 December 2018, https://www.alvaraalto.fi/en/services/seminars/alvar-aalto-researchers-network-seminar-2012/.

6   DOCOMOMO International Specialist Committee on Technology (ISC/T), *Technology of Sensations: Preservation Technology Dossier 7* (Copenhagen: The Alvar Aalto Vyborg Library, Royal Danish Academy of Fine Arts, 2004), pp. 76, 83, 91 and 100, accessed 30 December 2018, https://issuu.com/docomoisctechnology/docs/dossier_7_-_technology_of_sensation.

7   Maija Kairamo, Tapani Mustonen, *The Alvar Aalto Library in Vyborg*, DOCOMOMO International Specialist Committee on Technology (ISC/T), DOCOMOMO Preservation Technology Dossier 7: *Technology of Sensations: The Alvar Aalto Vyborg Library*, Royal Danish Academy of Fine Arts, Copenhagen, 2004, pp. 44–55.

8   For further and more detailed information on the restoration process of Viipuri Library, see: Erik Adlercreutz, et al. (eds.), *Alvar Aalto Library in Vyborg. Saving a Modern Masterpiece* (Helsinki: Rakennustieto, 2009), p. 144; and Erik Adlercreutz, Maija Kairamo, and Tapani Mustonen, (eds.), *Alvar Aalto Library in Vyborg: Saving a Modern Masterpiece. Part 2* (Helsinki: Rakennustieto, 2015), p. 127.

9   Europa Heritage Europa Nostra Awards, "Viipuri Library in Vyborg," last modified 14 April 2015, accessed 30 December 2018, http://www.europeanheritageawards.eu/winners/viipuri-library-vyborg/.

10 Sergej Kravchenko, "Restauración de la biblioteca de Viipuri," *Loggia* 4 (1997): pp. 30–31, accessed 30 December 2018, https://polipapers.upv.es/index.php/loggia/article/view/5373/5358.

11 A good overview of the restoration works is available at: "Restoration," accessed 15 December 2018, http://aalto.vbgcity.ru/node/366.

12 Alvar Aalto Foundation, "Viipuri Library Restoration Project Makes Progress," last modified 15 February 2011, https://www.alvaraalto.fi/en/news/viipuri-library-restoration-project-makes-progress/.

Fallingwater

Mill Run, Pennsylvania, USA
Frank Lloyd Wright, 1937

Fallingwater in Mill Run, Pennsylvania, 1985.

Original steel frame windows in living room, 1985.

Fallingwater, designed by American architect Frank Lloyd Wright (1867–1959), is one of the most renowned examples of Modern architecture in the US. Commissioned in 1935 by Edgar Kaufmann, the head of the Kaufmann's Department Store in Pittsburgh, it exemplifies Wright's idea of organic architecture, where the boundaries between exterior and interior are blurred, and the interiors are visually and metaphorically connected to the site. Completed in 1937, the house was expanded two years later with an addition, also designed by Wright, which included a guest house connected to the original building by a set of semicircular stepped concrete canopies. In his design for the Kaufmanns, Wright materialized his vision of organic architecture by including broad expanses of steel-framed fixed units, casement windows and doors built with hot-rolled Z- and T-shaped sections manufactured by Hope's Windows.[1] Four T-shaped vertical steel members, embedded during the original construction within the concrete parapet at the south edge of the living room, double as both window mullions and structural support to the cantilevered terrace above.[2]

The steel-framed glazed enclosures at Fallingwater, painted in Wright's characteristic Cherokee Red, not only help to define

Corroded and inoperable original steel frame windows at terrace, 1985.

Concrete damage and leaking steel frame windows, 1985.

intimate interior spaces, they also connect them to the exterior in unique ways. In the living room, a steel frame glazed hatch with overhead sliding panes opens to a stair leading down to the Mill Run Creek that runs past the house before cascading into a waterfall. On the opposite end of the living room and at each bedroom, the steel frame glazed enclosures open up to the terraces cantilevering over the waterfall. Along a multistory window wall parallel to the stone masonry chimney shaft, the west corners of the rooms vanish when their out-swing casements windows are opened. This nontraditional solution affords uninterrupted views towards the adjacent forest, bringing nature's sounds and cool moist air into each room. On the opposite end of the same window wall, the glass panes and transoms of the fixed unit do not terminate in a steel frame or a mullion but are embedded within the stonework of the chimney shaft. The fusion of these two assemblies blurs the boundaries of window and wall, of interior and exterior, and reinforces the openness and spatial fluidity that define Fallingwater's sense of place.

CONDITION PRIOR TO INTERVENTION

In 1963, the building was entrusted to the Western Pennsylvania Conservancy (WPC) with the mandate "to maintain in Fallingwater its character as a weekend home of the period it was occupied by its owners, Edgar and Liliane Kaufmann (occupied 1937 to 1955) and their son, Edgar Kaufmann, Jr. (occupied 1937 to 1963)."[3] Since then, the Main House and the majority of the Guest House have been open to the public as a museum. The former service quarters located on the upper level of the Guest House now serves as WPC administrative offices.

TIMELINE

1935	Designed
1937	Main house completed
1939	Guest house completed
1937–1963	Used as Kaufmann family home
1963	Donated to Western Pennsylvania Conservancy (WPC)
1964	Opened as public museum
1976	Listed as a National Historic Landmark
1989	Glass replacement work implemented including installation of single-pane glazing with UV-protective glazing
1992–2003	Glazing is removed, steel frames are restored
1995–2002	Structural assessment and post tensioning work designed by Robert Silman Associates, PC (now Silman)
2005	First phase of building intervention completed
2009	WPC launches the Window Legacy Fund
2009–present	Ongoing steel frame glazed enclosure conservation and glass replacement
2019	Designated a UNESCO World Heritage Site (with seven other buildings designed by Frank Lloyd Wright)

Steel frame window wall, 1985.

The early interventions on the glazed assemblies of this iconic historic house focused on repair and restoration of the glazed enclosures as originally designed.[4] The lack of adequate waterproofing details led to water infiltration into the house through the concrete terraces and the round drip edges along the roof perimeters. This condition resulted in corrosion of the steel-framed glazed enclosures, affecting everything from the fixed casement window mullions to the window wall at the kitchen. In 1989, as the original 1/4" (6mm) plate glass panes broke, they were replaced (in the Main House and guestroom portion of the Guest House, but not in the former service quarters) with single-pane laminated glazing with a UV-protective film by Saflex.[5] Glass failure continued to occur however, and the cause of the breakage in certain areas of the Main House was later determined to be the result of progressive deflection of the cantilevered terrace slabs.[6]

Restored steel frame windows (foreground) and window wall (beyond), 2016.

Restored corner windows at the Guest House from the exterior, 2016.

INTERVENTION

In the period between 1992 and 2003, glass panes were removed and original steel frames were cleaned of corrosion products down to bare metal; severely corroded sections were cut out and replaced in-kind. The hardware was also replaced and all operable units were refurbished to make them functional.[7] All of the steel frames were primed with Poly-Ura-Prime and finished with two coats of Endura-Shield Series 73, a high performance coating system by Tnemec. The single-glaze glass panes were reinstalled with new aluminum glazing beads and Vulkem 116 putty (by Mameco/Vulkem). This phased building-wide intervention was completed in 2005.[8] The window restoration work was performed in coordination with the post-tensioning retrofit of the sagging cantilevered reinforced-concrete terrace structures and with the waterproofing and water management solutions implemented on the roof, terraces and eaves to eliminate the leaks.[9]

View of restored steel frame terrace door and window enclosure, 2016.

Over a decade after the first restoration was completed, glass breakage and failure of the surface-mounted UV protection film led to another glass replacement campaign. To replace the glass, the WPC launched a new fundraising effort in 2009 known as the Window Legacy Fund. As the organization explained on their website, "Fallingwater's windows—signature components of this iconic house—are suffering from the effects of weather and time. The window glass must be replaced in order to protect the house, its furnishings and its priceless collections from heat, sunlight and UV radiation."[10] At the time of this writing, the glass panes

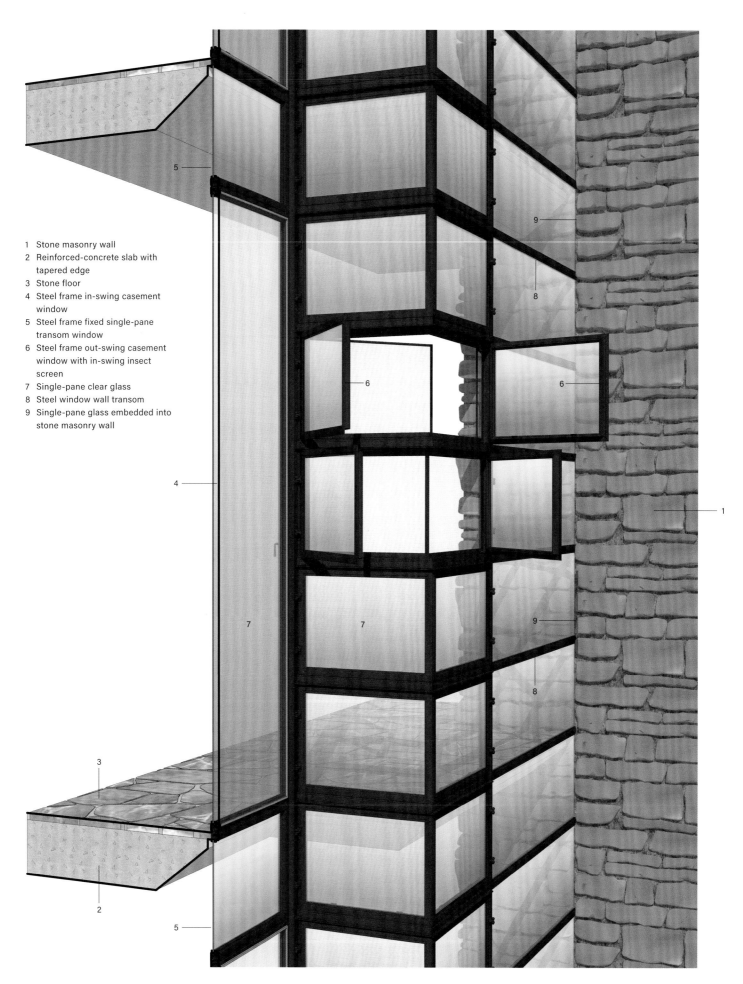

1 Stone masonry wall
2 Reinforced-concrete slab with tapered edge
3 Stone floor
4 Steel frame in-swing casement window
5 Steel frame fixed single-pane transom window
6 Steel frame out-swing casement window with in-swing insect screen
7 Single-pane clear glass
8 Steel window wall transom
9 Single-pane glass embedded into stone masonry wall

Typical section at existing steel frame window wall.

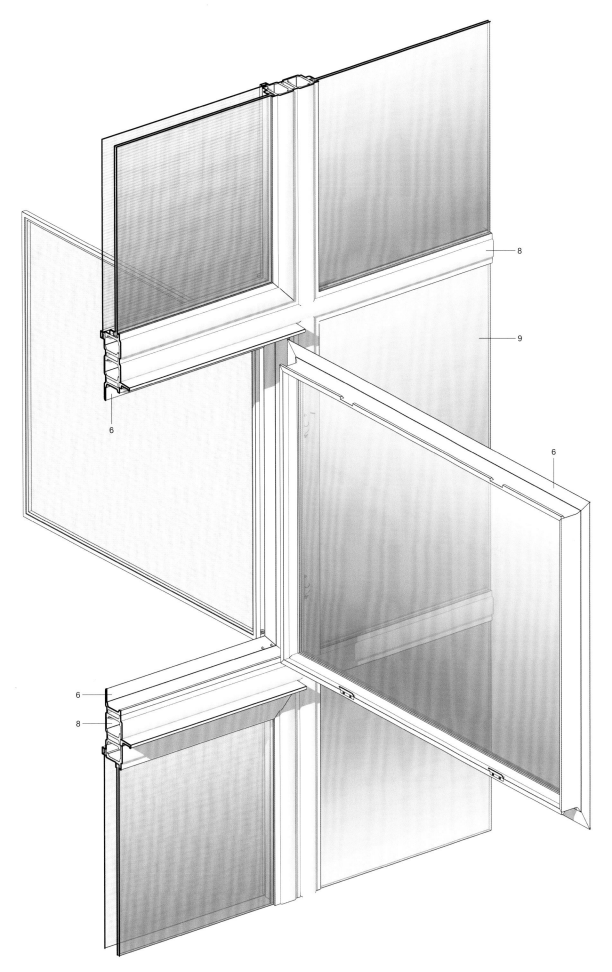

Typical window wall detail.

Frameless steel window wall jamb with glass panes embedded in the adjacent masonry wall, view from interior, 2016.

Close-up interior view of frameless steel window wall jamb with glass panes embedded in the masonry wall, 2016.

are in the process of being removed and replaced as funding allows. The ongoing project consists of the replacement of the existing single-pane glass with laminated units, a solution that was developed in 2011 after various test trials. In areas requiring UV protection, the replacement glass is comprised of two 1/8″ (3mm) thick sheets of PPG low-iron Starphire glass with an interlayer of Kuraray (formerly DuPont) SentryGlas. Use of the colorless SentryGlas interlayer eliminates the need for (and the problems associated with) an applied surface-mounted UV film. The final assembled dimension is just over 1/4″ (6mm) thick, allowing it to be fitted into the historic sash. Where no UV protection was necessary, tempered 1/4″ (6mm) PPG Starphire glass without the interlayer is being used.[11]

The glass replacement work is also being coordinated with ongoing steel conservation work at the windows. This is usually implemented in alternate years on areas that show signs of corrosion (typically near thresholds or, in the case of the glazed hatch,

Interior view of the window wall corner showing the restored out-swing corner casement windows with in-swing insect screens, 2016.

Restored steel frame out-swing corner casement windows from the exterior, 2016.

where they are near water or attract condensation). At these locations the work areas are cleaned, the paint is removed if necessary, the steel is cleaned and repairs are made if needed, and then the surfaces are repainted. Given that the steel conservation work requires removal of the glass, the WPC usually tries to select areas where the glass can be replaced during the steel conservation work. That includes windows and doors where the surface-mounted film has begun to delaminate (they start to fog around the perimeter), or has cracked.[12]

COMMENTS

The restoration of the original steel frame windows at Fallingwater was limited to localized in-kind steel replacement, cleaning and refinishing with a high performance coating and glass replacement with a surface-mounted UV protection film. This approach had the benefit of retaining historic fabric belonging to a given period of significance, while allowing for potential compromise (to a certain extent) in order to accommodate ongoing use and conservation of fragile interior content. Compared to the structural repairs at the cantilevered terraces, or to the exterior waterproofing work implemented as part of the restoration, the intervention at the steel windows and doors was more modest. It was appropriate, but relatively short-lived. Shortly after the building structure was stabilized, the leaks minimized and the steel frame assemblies restored, another intervention was required.

Exposure to the elements, and in particular to the highly-humidified microclimate around the waterfall, requires periodic maintenance to mitigate corrosion on the steel members. The humid site environment has proven that a surface-mounted UV

Sliding steel frame glazed hatch and sliding panes above the stairs leading down to Bear Run Creek, 2016.

CREDITS

Fallingwater (1937)
Mill Run, Pennsylvania

ORIGINAL CONSTRUCTION

Architect
Frank Lloyd Wright

Structural Engineers
William Wesley Peters and
Mendel Glickman

1992–2005 INTERVENTIONS

Architects
Wank Adams Slavin
Associates, LLP

Owner
Western Pennsylvania
Conservancy

Materials Conservator
Integrated Conservation
Resources, ICR

Structural Engineers
Robert Silman Associates, PC

Conservator
Norman Weiss, FAPT

**Window Restoration
Contractor**
Seekircher Window Repair

View of restored steel frame windows from inside the Guest House, 2016.

protection film application is just not durable. The proposed replacement with laminated glass is a more suitable option to ensure long-lasting UV protection. However, the challenge posed by the site's environmental conditions may require a more intrusive intervention on the steel frames within a few years, such as temporary coating removal and reinstallation after onsite metalizing of the steel frames, or perhaps an off-site solution consisting of hot-dip galvanizing after temporary removal and salvage. The steel frame glazed enclosures at Fallingwater are a unique example of the challenges posed by Modern buildings, particularly in regards to historic house museums and in locations with more extreme site conditions. The WPC's conservation-focused management approach for Fallingwater is an interesting case study on heritage conservation decisions aimed at retaining authenticity of place and materials while balancing durability, cost effectiveness and a long-term vision for the historic site.

NOTES

1   Norman Weiss, Pamela Jerome, and Stephen Gottlieb, "Fallingwater Part 1: Materials-Conservation Efforts at Frank Lloyd Wright's Masterpiece," *APT Bulletin* 32, no. 4 (2001): p. 48.

2   Robert Silman, "The Plan to Save Fallingwater," *Scientific American* (September 2000), p. 93.

3   Western Pennsylvania Conservancy, "Mission," accessed 12 October 2015, http://www.fallingwater.org/64/.

4   Pamela Jerome, "Restoration and Replication of Steel Elements at Fallingwater and Solomon R. Guggenheim Museum," National Center for Preservation Technology video, 20:27, posted 22 July 2015, accessed 16 December, 2018, https://ncptt.nps.gov/blog/restoration-and-replication-of-steel-elements-at-frank-lloyd-wrights-fallingwater-and-solomon-r-guggenheim-museum/.

5   Weiss, Jerome, and Gottlieb, "Falling Water Part 1," pp. 52, 55.

6   Silman, "The Plan to Save Fallingwater," p. 94.

7   Weiss, Jerome, and Gottlieb, "Falling Water Part 1," p. 52.

8   Pamela Jerome, Norman Weiss, and Hazel Ephron, "Fallingwater Part 2: Materials-Conservation Efforts at Frank Lloyd Wright's Masterpiece," *APT Bulletin* 37, no. 2/3 (2006): pp. 8–9.

9   For more information on the structural retrofit, see: Matthew L. Waldsept, "Rescuing a World-Famous but Fragile House," *New York Times*, 2 September 2001, https://www.nytimes.com/2001/09/02/us/rescuing-a-world-famous-but-fragile-house.html. For more information on the waterproofing and water management solutions implemented, see: Jerome, Weiss, and Ephron, "Fallingwater Part 2: Materials-Conservation Efforts at Frank Lloyd Wright's Masterpiece," *APT Bulletin* 37, no. 2/3 (2006): pp. 3–11.

10 Western Pennsylvania Conservancy, "The Fallingwater Window Legacy Fund," accessed 16 December 2018, https://www.fallingwater.org/donate/more-ways-to-give/windows-legacy-fund/.

11 Scott Perkins, email message to Angel Ayón, 22 March 2016.

12 Ibid.

Farnsworth House

Plano, Illinois, USA
Ludwig Mies van der Rohe, 1951

Interior view of the Farnsworth House in Plano, Illinois, looking out onto the terrace from the south facade, undated image.

Exterior view of the south facade and terrace, undated image.

Dr. Edith Farnsworth, a single independent professional, commissioned the house from German-American architect Ludwig Mies van der Rohe (1886–1969) as a weekend retreat outside Chicago. Designed between 1945 and 1946 and built between 1949 and 1951, the Farnsworth House represents Mies' vision of Modern architecture as "almost nothing", which stems from his ideal to reduce "every element to its essence."[1] The result is a combination of simplicity and geometric refinement creating a unique space within a minimal framework of translucent "skin and bones." Through this transparent skin, the spare interior merges with the expansive exterior, becoming one unified experience. According to Mies, "when one looks at Nature through the glass walls of the Farnsworth House, it takes on a deeper significance than when one stands outside. More of Nature is thus expressed—it becomes part of a greater whole."[2]

Unidentified woman (presumably Edith Farnsworth) at the Farnsworth House, undated image.

The house consists of two adjacent rectangular structures at different levels raised above the ground by steel columns. The lower structure is a steel frame open terrace with travertine floor. A set of steel frame steps with travertine threads leads from the open terrace up to a covered terrace at the upper structure and another set of steps leads down to the ground. The upper structure also has travertine floors at both the covered terrace and the interior space enclosed by a steel frame window wall. The interior space is organized around a central core that separates the entrance to the west, the living space to the south, the kitchen to the north, and the bathroom and bedroom to the east, which is the

North facade with insect screens installed by Edith Farnsworth that were later removed, 1971.

only location where the steel frame window wall has operable (hopper) windows.

As stated in its National Historic Landmark nomination, the architecture of the Farnsworth House "represents an extreme refinement of Mies van der Rohe's minimalist expression of structure and space."[3] The "skin" of the house is defined by the floor-to-ceiling perimeter window wall, which was originally constructed with 1/4" (6mm) single-glazed polished plate glass. Unfortunately, Dr. Farnsworth was not at ease with the cost and design of the house and openly complained about it being too "transparent, like an X-ray."[4] Upon completion of the house in the summer of 1951, she barred Mies from selecting the furniture for the house and also pressed for the installation of a wood-framed insect screen around the covered terrace that had been part of the original 1947 design, but which was presumably not favored by Mies.[5]

CONDITION PRIOR TO INTERVENTION

The house was still occupied by Dr. Farnsworth in 1954 when it flooded for the first time and the water rose 24" (61cm) above the floor.[6] Flooding episodes continued to take place during Dr. Farnsworth's occupancy of the house and after the property was purchased by Lord Peter Palumbo in 1971 (who, incidentally, removed the screens installed around the covered terrace by Dr. Farnsworth). In 1996, an unprecedented two-day rainstorm deposited 17" (43cm) of water on the site, causing a wave of silt and mud that shattered one of the original plate glass windows and cov-

TIMELINE

1945–1946	Design phase
1949–1951	Construction phase
1954	First flood
1971	Lord Peter Palumbo buys house. Removes insect screens that Farnsworth installed around the upper terrace
1996	Fox River rises 60" (1.5m) cresting 17" (43cm) above the floor; glass is broken, furniture and artworks swept away
2003	Owned and operated as a house museum by National Trust for Historic Preservation
2004	Listed in the National Register of Historic Places
2006	Designated as National Historic Landmark
2008	Flooded by remnant rains of Hurricane Ike
2012–2013	Investigation and trial repairs by Krueck and Sexton Architects and WJE

View of the site looking toward the Farnsworth House (right) and the Fox River (left), 2016.

View of the north facade during a severe flood event in 2008, where water rose well above the interior floor level.

ered the travertine floors.[7] Another flood took place in 1997, when the water rose up to 2″ (5cm) above the floor. Lord Palumbo eventually sold the house to the National Trust for Historic Preservation in 2003, which has since been the steward of the house and the site. The house is currently within both the 100-year and 500-year local flood plains. With the effects of climate change and ongoing urban development upstream of the Fox River, the threat of flood continues. Subsequent to the 1996 event, flooding has affected the property in 1997, 2008 and twice in 2013.[8]

Over the years, in order to address the adverse effects of flooding and accidental breaking, the glass was replaced several times, including some of the larger 10′ (3.05m) wide single-glazed panes. Of the seventeen window bays, only six original polished plates of glass (or early replacement plate glass) remained, as of 2013. Eight panes had been replaced with float glass (installed sometime after 1959) and the remaining three panes, as well as the glass on the original aluminum frame entrance doors (manufactured by Kawneer), had been replaced with tempered glass (installed at some point after the 1970s).[9] The combination of original plate glass and both float and tempered replacement led to concerns about the safety and appearance of the glass throughout the house. For instance, the use of original annealed glass and other non-tempered replacements is now in conflict with contemporary safety requirements that require the installation of safety glass when glazing height is within a limited distance from the finished floor. In order to address these safety concerns, a safety film was applied on all of the non-tempered glass, including both the original and the replacement polished plate glass that was installed after 1959. While the film provides the code-compliant safety protection required to open the house to the public, the width of each film roll is narrow and vertical seams were needed to cover each bay, roughly 10′ (3.05m) x 10′ (3.05m). Unfortunately,

Farnsworth House. Exterior view of the south facade and terrace, 2016.

Farnsworth House. Steel frame single-glazed window wall at the south facade, 2016.

Exterior view of the south window wall pointing to the seam between the surface-mounted safety films installed on the interior side, 2016.

Close-up of the replacement single-pane tempered glass installed in the original steel frame at the south facade, 2016.

the non-original replacement tempered glass panes can be easily identified from the exterior under specific lighting conditions (bright exterior, darker interior with clear sky). While walking around the house the reflections on the glass are distorted as a result of the roller-wave tempering process. Furthermore, all of the existing tempered glass has the ANSI-required labeling in one corner, including the entrance doors.

In addition to ongoing glass breakage, there has been noticeable corrosion of the steel frames, particularly at the sills adjacent to the travertine floor pavers. As a result of rust jacking, an adverse effect created by the expansion of corroding steel, the steel frames have warped, increasing stresses on the glass.

INTERVENTIONS

During the 2012–2013 intervention, led by Chicago-based Krueck and Sexton Architects and Wiss Janney Elstner, the project team considered various reglazing options. The most conservative approach included installing float glass with a safety film, similar to that existing at other locations. This approach, however, was unlikely to meet current building code requirements regarding impact safety. Another approach considered temporary removal of the original steel glazing stops (fixed in place with exposed countersunk fasteners installed in pre-drilled holes) and enlarging the existing glazing pocket to accommodate the face clearance required by the new code-compliant 1/2″ (12.7mm) thick, replacement tempered glass or 5/8″ (16mm) thick replacement laminated, annealed glass. However, this approach would have required relocating the exterior glazing beads roughly 1/8″ (3mm).[10] This would have resulted in unevenness of the exterior details, or would have required a more aggressive intervention to replace all the glass, including the remaining original panes, to achieve a uniform glass plane from both the interior and the exterior.

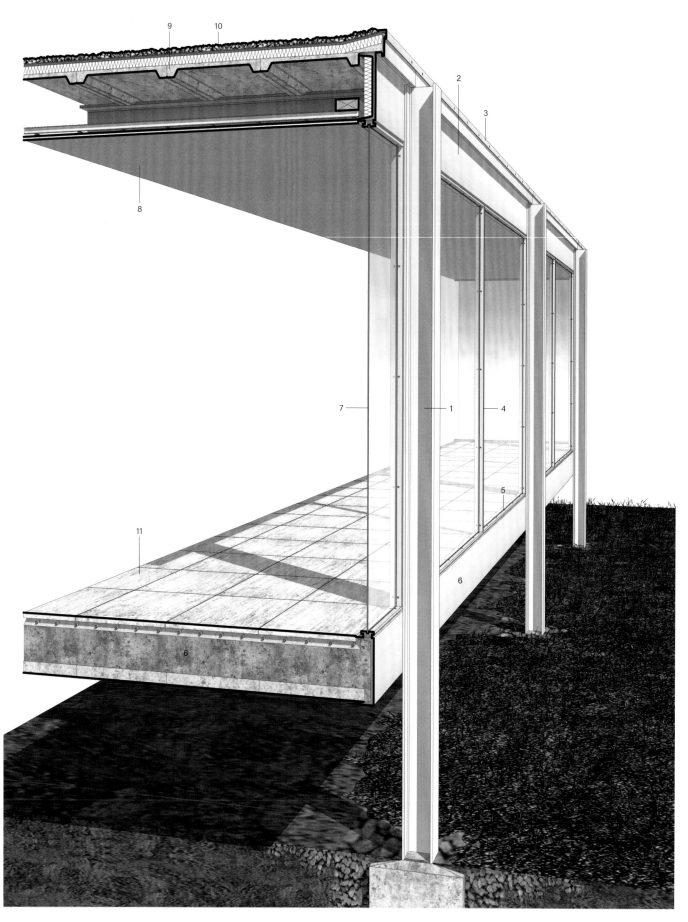

Typical section at existing steel frame window wall.

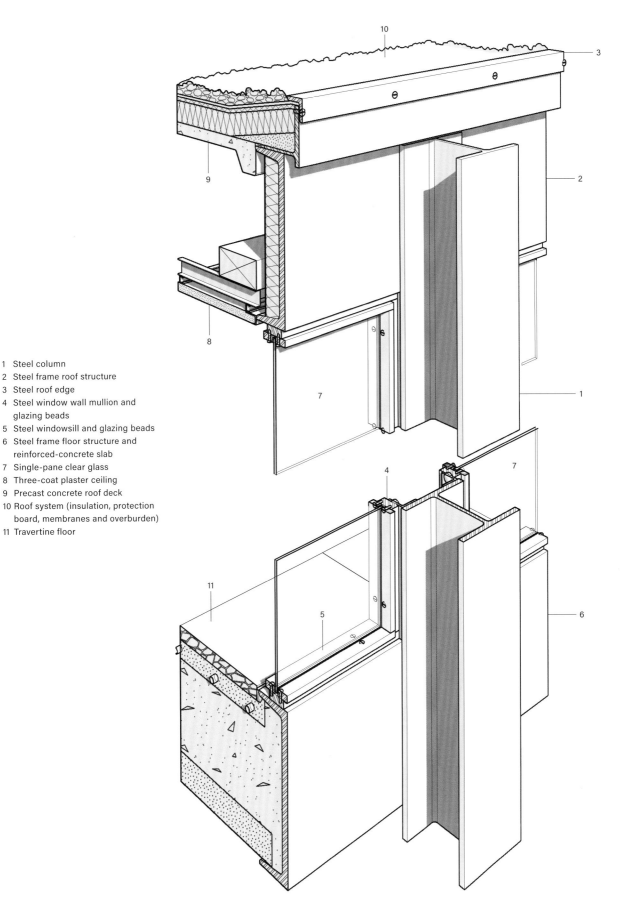

1 Steel column
2 Steel frame roof structure
3 Steel roof edge
4 Steel window wall mullion and
   glazing beads
5 Steel windowsill and glazing beads
6 Steel frame floor structure and
   reinforced-concrete slab
7 Single-pane clear glass
8 Three-coat plaster ceiling
9 Precast concrete roof deck
10 Roof system (insulation, protection
   board, membranes and overburden)
11 Travertine floor

Typical window wall details.

Interior view looking out of the south facade towards the Fox River, 2016.

To avoid these unfavorable results, the window wall renovation used replacement 1/4″ (6mm) clear, tempered heat-soaked glass with 3/1000″ (0.076mm) maximum peak to valley roller-wave distortion. The glass specification also called for surface #1 (bottom/base edge) to be parallel to the tempering oven rollers, resulting in a horizontal roller-wave orientation on the installed glass. The design team also called for 8/1000″ (0.2mm) maximum edge curl, a tolerance for localized warp of 1/32″ (0.8mm) over any 12″ (0.3m) span and a tolerance for overall bow and warp of 1/32″ (0.8mm) per lineal foot (0.3m).[11] According to the National Trust for Historic Preservation, the specifications for the replacement glass were aimed at achieving surfaces as smooth as possible and with a thickness matching the originals so that the framing can be reused and avoid altering the original design and detailing.[12] The work also included repairing all of the sill frames, replacing the sill stops, and refinishing using the paint specified when Palumbo restored the house.[13]

In 2015, the National Trust elected to replace three large glass panes and two panes on the hopper windows with the same specification of 1/4″ (6mm) thick tempered glass. Although this solution is a compromise, the Trust expects that it will continue to be implemented when glass replacement is required until a better one can be found that maintains the original material and detailing of the windows.[14]

Exterior view of corrosion along the sill of the west window wall, 2016.

Close-up interior view of the corroding steel window frame and seam between surface-mounted safely film applications, 2016.

Corrosion and rust jacking deformation along the bottom of the west window wall, 2016.

Interior view looking east, 2016.

View from the interior looking southwest towards the Fox River, 2016.

For the National Trust, a long-term solution to preventing rust jacking is paramount, which means securing the house against flooding.[15] Many flood mitigation options have been investigated and considered for the Farnsworth House. These range from filling the site and raising the house to a higher plane, relocating the house to higher ground or installing underground hydraulic lifts to elevate the house when required, all the way to a buoyant amphibious foundation that relies on natural buoyancy to lift it above flood waters, or installing a circular below-grade flood-barrier around the house that could be raised to protect it during a flooding event.[16] Currently, the National Trust is proceeding with the design of a hydraulic system that will lift the house above the flood waters and then lower it back down once the waters have receded.[17] Once the system is in place, the Trust plans to perform the deferred comprehensive restoration and repair work that the house requires.

COMMENTS

The last interventions at the Farnsworth House have struck a fine balance between serviceability, concerns for safety, preservation of historic fabric and design intent. For a house of this significance where the interior space, outdoor views and exterior enclosure are

Steel frame detail at the terrace and south facade, 2016.　　　East facade steel frame detail, 2016.

mostly defined by the perimeter glazing, compromising the views—even if slightly—while retaining the original historic fabric and design intent was rightly deemed inappropriate. Monolithic tempered glass, the current preferred glass replacement option, has less lateral load resistance than laminated glass, but fits within the existing glazing pockets and represents a less invasive approach. However, the visual distortions associated with tempered glass, even if minimized in response to the specification requirements, are a concern to be taken into account, particularly for a cultural resource like the Farnsworth House where the glass is a significant, character-defining feature.

At the Farnsworth House, implementation of a viable flood prevention plan is just as important as the criteria for glass replacement. Flooding adversely affects the furniture, interior fea-

CREDITS

Farnsworth House (1951)
Plano, Illinois

ORIGINAL CONSTRUCTION

Architect
Ludwig Mies van der Rohe

1971–2015 INTERVENTIONS

Owner
Lord Peter Palumbo
(1971–2003)

National Trust for Historic
Preservation (2003-present)

Historical Architects
Kreuck and Sexton Architects

Exterior Envelope Specialists
Wiss Janney Elstner
Associates, Inc. (WJE)

Restoration Contractor
Berglund Construction

**Metal Work and Glazing
Subcontractor**
MTH/Illinois Bronze and Metal

Exterior view looking northeast of the house and its terraces, 2016.

West facade, 2016.

NOTES

1   Historic American Buildings Survey, *Edith Farnsworth House,* (Washington, DC: US Dept. of the Interior, National Park Service, Cultural Resources, Historic American Buildings Survey/Historic American Engineering Record, 2010), HABS No. IL-1105 (Addendum to), p. 9.

2   Adelyn Perez, "AD Classics: The Farnsworth House/Mies van der Rohe," *ArchDaily,* last modified 13 May 2010, https://www.archdaily.com/59719/ad-classics-the-farnsworth-house-mies-van-der-rohe/.

3   Anthony Raynsford, *National Historic Landmark Nomination: Farnsworth House* (Washington, DC: US Department of the Interior, National Park Service, September 2004), p. 4.

4   Peter Blake, *Mies van der Rohe, Architecture and Structure* (Baltimore: Penguin Books, 1964), pp. 85–89, accessed 28 December 2018, http://www.columbia.edu/cu/gsapp/BT/GATE-WAY/FARNSWTH/blake.html; Elizabeth Gordon, "The Threat to the Next America," *House Beautiful* 95 (April 1953): p. 129.

5   The relationship between Dr. Farnsworth and Mies was quite contentious, and many presumed their relationship went beyond that of client-architect. See: Nora Wendl, "Sex and Real Estate, Reconsidered: What Was the True Story Behind Mies van der Rohe's Farnsworth House?" *ArchDaily,* accessed 29 May 2016, http://www.archdaily.com/769632/sex-and-real-estate-reconsidered-what-was-the-true-story-behind-mies-van-der-rohes-farnsworth-house.

6   Ashley Wilson, "After the Storm: Glazing Replacement Issues at the Farnsworth House," paper presented at APT/DOCOMOMO DC Chapter, fall 2014 *Symposium: The Challenges of Preserving Modern Materials & Assemblies,* 26 September 2014.

7   Ben Schulman, "Shoring Up Is Hard To Do," *Architect* 104, no. 12 (December 2015): p. 40.

8   Farnsworth House, "Flood Mitigation Project – History of Flooding at Farnsworth House," accessed 16 December 2018, https://farnsworthhouse.org/history-of-flooding/.

9   Wilson, "After the Storm: Glazing Replacement Issues at the Farnsworth House."

10 Ibid.

11 Ibid.

12 Maurice Parrish, e-mail message to author, 24 May 2016.

13 Ibid.

14 Maurice Parrish, email message to author, 26 May 2016.

15 Ibid.

16 Farnsworth House, "Flood Mitigation Project," accessed 16 December 2018, https://farnsworthhouse.org/flood-mitigation-project/. Detailed reports of the flood mitigation options explored are available for download at: Farnsworth House, "Portfolio," accessed 28 December, 2018, https://farnsworthhouse.org/portfolio-items.

17 Maurice Parrish, email message to author, 24 May 2016.

tures and the glass panes, and also increases the corrosion rate of the steelwork in general. The windowsills and glass stops at the house have been submerged in floodwater for days at time, on more than one occasion, and are more corroded than the rest of the structure. In this case, flood prevention is a protection measure for the window wall steel frames and glass—unprecedented, yet a necessary implementation when the alternative would be losing the historically significant original glazed assemblies. This unique intervention would raise the house above the flood plain temporarily as needed to protect it from new flooding. Flood prevention alone, however, will not prevent or decelerate ongoing corrosion-related damage at the Farnsworth House steel frame window wall. A repair campaign is needed to temporarily remove, salvage and reinstall the glass to allow for steel cleaning, repair of damaged steel, and painting. Ongoing maintenance will also be required, whether the preferred hydraulic lift flood-prevention solution or other suitable measure is implemented. The challenge in restoring and maintaining the Farnsworth House lies in balancing cultural significance with preservation standards and innovating approaches that are both authentic and appropriate.

TWA Flight Center

JFK International Airport, Queens, New York, USA
Eero Saarinen, 1962

West facade and main entrance to the TWA Flight Center at Idlewild Airport (now JFK) in Queens, New York, ca. 1962. © Ezra Stoller/Esto

The Trans World Airlines (TWA) Flight Center at New York International Airport in Queens (now known as John F. Kennedy Airport) is the last and probably best-known and most influential project by renowned Finnish-American architect Eero Saarinen (1910–1961). For the TWA terminal, Saarinen was interested in interpreting "the excitement of air travel" and in revealing the terminal "not as a static enclosed space but as a place of movement and transition."[1] He transcribed these aspirations into a unique, welcoming space defined by four soaring thin-shell concrete vaults supported by, and cantilevered off, four pairs of sculptural concrete columns. The west shell is the smallest and faces the entrance. The east shell is the widest and faces the tarmac. The other two symmetrical shells extend above the two-story north and south wings. Linear aluminum-framed skylights between the four vaults provide daylighting and create a continuum between concrete shells and piers. Curved and tilted lock-strip window walls rise up and out to the exterior perimeter of the shells. The window walls have monolithic 1/4" (6mm) plate glass secured

Interior view of the sunken lounge and adjacent steel frame window wall, ca. 1970.

Tilted lock-strip steel frame window wall with bow-shaped steel trusses in the departure lounge, ca. 1976.

Interior view of the ticket counter and Solari board adjacent to west window wall, ca. 1970.

with lock-strip neoprene gaskets supported by aluminum extrusions anchored to vertical, bow-shaped steel trusses that act as mullions and to horizontal stock steel bars that perform as transoms. These "dramatically slanting arched windows convey a powerful image of thrust."[2]

At the TWA Flight Center, the tilted steel frame window walls create a glazed enclosure that is as integral a part of the building design and morphology as its swooping curvilinear profile. The translucency of the window wall allows for exterior views in multiple directions, and reinforces the illusion of weightlessness and imminent take-off that characterizes Saarinen's design for the terminal. They are truly gateways to a metaphoric flight, prompting architect Robert A. M. Stern to call the TWA Flight Center the "Grand Central of the jet age."[3]

The original terminal, the south flight wing and the connecting tube opened in 1962. The north flight wing, its connecting tube and an underground baggage tunnel opened in 1970. With the continuing growth of the aviation industry, the terminal required

TIMELINE

1956–1959	Design period
1959–1962	Construction period
1962	Main terminal, Flight Tube 2 and Flight Wing 2 open
1967	Flight Tube 1 and Flight Wing 1 constructed by Saarinen's successor firm
1970	Underground passenger access tunnel to Flight Wing 1 added
1994	Designated a New York City Landmark
2005	Listed on the National Register of Historic Places
2005	Non-original additions demolished
2009–2012	First phase of restoration work completed
2016–2019	Construction of the new hotel and adaptive reuse of the former terminal

Section of tilted steel frame window wall with surface-mounted film in the departure lounge before restoration, 2009.

Tilted steel frame window wall in the departure lounge before restoration, 2009.

various additions and alterations to accommodate an ever-growing stream of passengers. Overcrowding, undersized public spaces, poor space security and a relentlessly inadaptable structural enclosure meant the building was always on the verge of being obsolete. By 1990, a canopy had been added by the entrance and another one by the baggage tunnel. The original sunken lounge facing the tarmac was removed in the early 1990s and additional ticket counters were added. Other additions ensued. The window wall was altered to accommodate new luggage tunnels. After various periods in and out of bankruptcy, TWA was finally purchased in 2001 by American Airlines. They held the lease on the building for another year, making additional alterations to the terminal, but ultimately did not renew it as the historic terminal was deemed unsuitable to contemporary flight requirements. The last flight departed from the TWA Flight Center in December 2001. The next year the building was officially closed and abandoned as a terminal.[4]

Subsequently, the building was designated as a local landmark, which prevented it from being demolished. The Port Authority of New York and New Jersey (PANYNJ), the airport operator, then stabilized the building and upgraded the mechanical systems in preparation for its future use as was required by the designation. For years, PANYNJ worked with other airlines, including Jetblue, to reuse the building. Construction of a new terminal for Jetblue to the east of the TWA Flight Center required removal of the original flight wings at the tip of the tubes, triggering a mandatory review based on Section 106 of the National Historic Preservation Act, which is aimed at considering the effects of federally funded projects on historic properties. On completion of the Section 106 review, the New York State Historic Preservation Office (SHPO) deemed the proposed alteration "an adverse effect" and imposed a series of mitigation measures to compensate for the loss of original historic fabric. This included nomination for listing on the National Register of Historic Places, despite the building being almost 20 years short of meeting the 50-year age threshold for inclusion on the register. It also resulted in a thorough documentation being completed, based on the Department of the Interior's Historic American Building Survey standards. An interior restoration was then undertaken and an interpretive exhibition about the history and development of JFK Airport and the TWA terminal was commissioned. Altogether, various stakeholders, including several local, regional and national historic preservation organizations, aviation experts and even the Finnish consulate, came together to form a new oversight entity known as the Redevelopment Advisory Committee, which continued to be active through the recently completed development.[5]

CONDITION PRIOR TO INTERVENTION

The first phase of the intervention on the TWA Flight Center included improvements to life safety, waterproofing repairs and restoration of interior finishes.[6] By 2005, an examination of the glazed enclosures revealed that the existing H-shaped lock-strip gaskets at the single-glazed window walls and ribbon skylights were at the end of their service life, which compromised the integrity and weather-tightness of the exterior enclosures. The original single-glazed system with segmented lock-strip zipper gaskets had deteriorated over the years, leaving the gaskets deformed and shifted, with open joints at most corner miters. Dark purple mylar UV films that had been installed on the interior glass surfaces during the 1990s in an effort to reduce glare and heat gain were not only unsightly, but also compromised visibility into the interior spaces. Several glass panes had broken and some had been replaced with plywood.

INTERVENTION

Beyer Blinder Belle Architects & Planners (BBB) led the historic research that guided the proposed restoration work for the TWA Flight Center. Their project team identified 1962–1970 as the period of significance to which the terminal should be restored. The re-

Visible daylight seen through failed original lock-strip gaskets at the window walls before restoration, 2009.

Poor repairs at the mitered joints of the lock-strip gaskets and corroding aluminum extrusion at the window wall before restoration, 2009.

Open joints at the steel frame window wall before restoration, 2009.

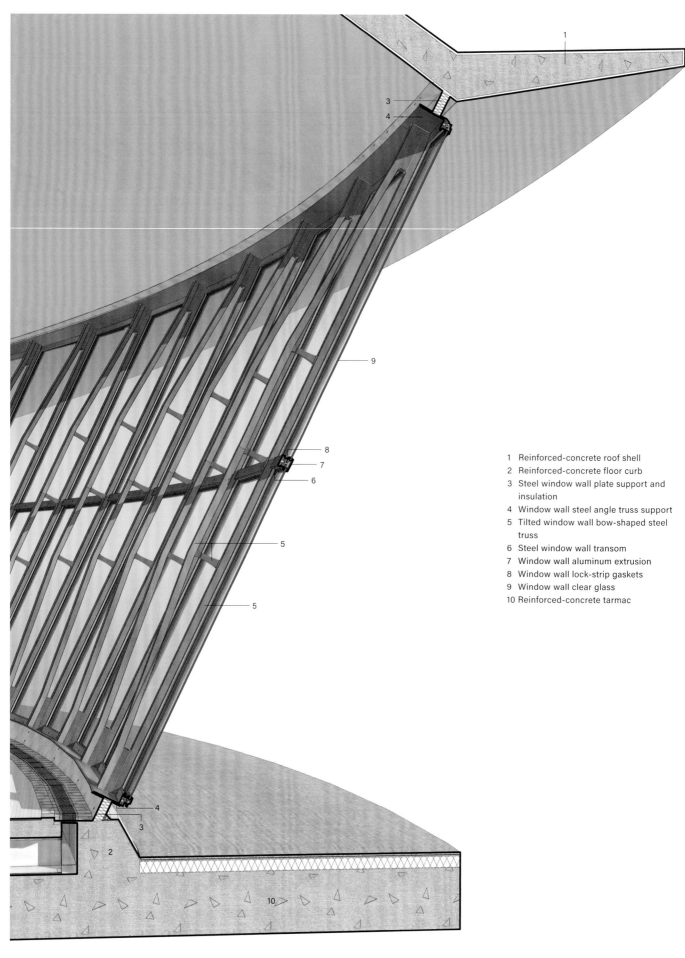

1 Reinforced-concrete roof shell
2 Reinforced-concrete floor curb
3 Steel window wall plate support and insulation
4 Window wall steel angle truss support
5 Tilted window wall bow-shaped steel truss
6 Steel window wall transom
7 Window wall aluminum extrusion
8 Window wall lock-strip gaskets
9 Window wall clear glass
10 Reinforced-concrete tarmac

Typical section at existing steel frame window wall.

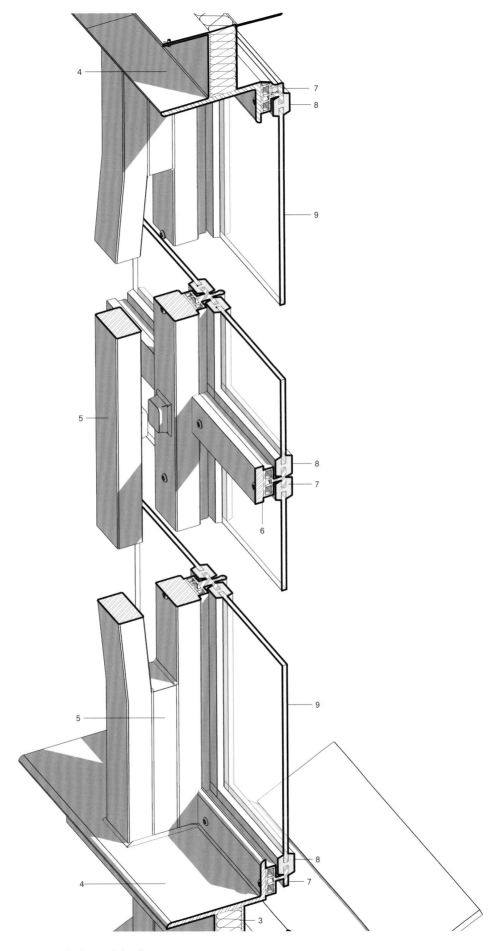

Typical window wall details.

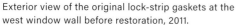

Exterior view of the original lock-strip gaskets at the west window wall before restoration, 2011.

Interior view of the north window wall before restoration and hotel addition, 2011.

sulting work, led by BBB and partially completed in 2012, included: removal of unsightly additions on the wings and east facade; concrete coating removal, concrete restoration and installation of new breathable coatings; abatement and restoration of interior finishes; removal of glass films; and glass cleaning.[7] At the unique steel frame sloped and curved window wall, the intervention included an analysis of the structural capacity of the existing steel trusses supporting the window wall. This analysis determined that replacing the existing single-pane glass with IGUs would have required reinforcing the steel trusses, thereby modifying the interior appearance of the window wall—an approach deemed inappropriate. As a result, the scope of work included removal of the existing glass, repair and refinishing the steel trusses and installing replacement 3/8″ (9mm) thick tempered glass by Viracon with contemporary unitized lock-strip gaskets matching the original design, sightlines and details.[8] The new neoprene extrusions for the replacement zipper gaskets were manufactured by the original fabricator based in Wisconsin, which was still in business. The unitized zipper gaskets reduced the number of seams and potential open joints at the miters, thus ensuring a more durable, reliable and watertight system. In addition to the repair of the existing steel frame trusses and replacement of the lock-strip single-glazed enclosure, the soaring concrete shells were repaired and recoated to ensure watertightness. Unfitting exterior additions were removed. The original distinctive sunken lounge opposite to the main entrance, which had been overbuilt and flattened, was also reinstated.

Interior view of the restored east and south window walls, 2015.

Interior view showing the restored east and west window walls, 2015.

Hotel additions under construction behind (east of) the former terminal, 2018.

COMMENTS

The TWA Flight Center is one of the most celebrated buildings of the 20th century worldwide. Its location at JFK Airport, one of the busiest international destinations in the US, exposes it to more than 59 million people each year.[9] Repurposing the obsolete terminal for a new use not only preserves the integrity of the building, but also ensures that its Modern legacy will be enjoyed by generations to come. The redevelopment project, completed in 2019, converted the former terminal into a gateway for a new hotel with a subterranean conference center, and added two symmetrical hotel wings between the original building and the Jetblue terminal behind.

CREDITS

TWA Flight Center (1962)
JFK International Airport,
Queens, New York

ORIGINAL CONSTRUCTION

Architects
Eero Saarinen
Kevin Roche and Cesar Pelli
(Associates)
Ralph Yeakel
(Supervising Architect)

Engineers
Abba Tor, Amman and Whitney

Mechanical Engineers
Jaros, Baum and Bolles

Lighting Consultant
Stanley McCandless

Acoustical Consultants
Bolt, Beranek & Newman

Contractor
Grove, Shepherd, Wilson &
Kruge, Inc.

2005-2009 INTERVENTIONS

Architects
Kevin Roche and John
Dinkeloo

2009-2019 INTERVENTIONS

Clients
Port Authority of New York &
New Jersey
TWA Redevelopment Advisory
Committee
MCR and MORSE Develop-
ment (Redevelopment)

Architects
Beyer Blinder Belle Architects
& Planners

**Redevelopment Consulting
Architect**
Lubrano Ciavarra Architects,
PLLC

Building Envelope Consultant
Gordon H. Smith Corporation

Structural Engineer
Arup

MEP Engineers
Jaros, Baum and Bolles

Lighting Design
SBLD Studio

**Graphics/Signage/Wayfind-
ing**
Pentagram

Fire Protection Engineers
Arup

Materials Conservator
Integrated Conservation
Resources, Inc. (ICR)

Cost Estimator
ELLANA Inc.

Landscape Architects
Matthews Nielsen Landscape
Architects

Salvaging the building by adaptively reusing it and restoring the steel frame window wall to its original configuration is a sensitive and appropriate intervention. Foregoing the temptation to upgrade the extensive original exterior glazed enclosures with more energy-efficient IGUs (which would have improved energy performance but noticeably altered the historic appearance of the terminal and compromised its sense of place) had a positive effect on maintaining the building's historic character, albeit a potentially negative one on the natural environment. The fact that the new TWA Flight Center hotel and conference center is aiming to be certified LEED Gold according to the rating system set forth by United States Green Building Council (USGBC) indicates that retaining the original single-glazed configuration is not an impediment for achieving sustainability goals.

The most challenging aspect of the redevelopment is the proposed hotel addition, and how it further alters the immediate context of the former terminal. The layout and integration of the new wings in relation to the historic building seem appropriate, but in the opinion of many in the New York preservation community, the height is not.[10] The new wings are slightly higher than the top of Saarinen's soaring concrete bird and feel too close and suffocating, even when seen from the front of the building, where they are less noticeable. The experience of the space from the restored sunken lounge and restored east window wall continues to be flanked by the original connecting tubes as it was originally. The most concerning and undesirable changes manifest more clearly at the north and south window walls adjacent to the upper lounges. The restored curved and tilted window walls now face the hotel addition and its dull, contemporary curtain wall. The hotel additions reduce the existing sky exposure and alter the visual relationships from within the upper-level lounges towards the overall airport context, replacing it instead with a view towards hotel rooms. This change redefines the original and current experience of place from the upper-level interiors. It also introduces an unintended shift of focus from the expansive possibilities of skyward travel to the mundane practicalities of a longer stay in an international airport. The "excitement of air travel" and sense of the terminal "not as a static enclosed space but as a place for movement and transition" embedded in the original design intent is compromised in the outward views through the restored steel frame windows walls. Unfortunately, this type of contextual change and experiential shift is one that cannot be overcome by material permanence and detailing alone.

NOTES

1 "Saarinen's TWA Flight Center," *Architectural Record* 132, no. 1 (July 1962): p. 129.
2 Herbert Mushamp, "Stay of Execution for a Dazzling Airline Terminal," *New York Times*, 6 November 1994, H31.
3 Ibid.
4 Richard W. Southwick. "Labor, Literature and Landmark Lecture Series: Life, Death and Rebirth of the TWA Flight Center," YouTube video, 1:19:02, posted 25 May 2016, accessed 21 December 2019, https://www.youtube.com/watch?v=84xg1dSGMwE.
5 Ibid.
6 Ibid.
7 Ibid.
8 Information on the replacement glass was provided by Richard W. Southwick, FAIA, LEED AP from BBB in person on 20 September 2016.
9 Bureau of Transportation Statistics, United States Department of Transportation, "Top 25 U.S. Freight Gateways, Ranked by Value of Shipments: 2008, 2009," retrieved 30 August 2015, https://www.bts.gov/archive/publications/americas_freight_transportation_gateways/2009/introduction_and_overview/figure_02; Airports Council International, "Passenger Traffic 2016 Final (Annual)," accessed 20 December 2018, https://aci.aero/data-centre/annual-traffic-data/passengers/2016-final-summary/.
10 David W. Dunlap, "Hotel Project Would Revive Embodiment of Jet Age at Kennedy Airport." *New York Times*, 6 December 2016, accessed 21 December 2019, https://www.nytimes.com/2016/12/06/nyregion/hotel-project-jet-age-kennedy-airport.html.

Zeche Zollverein

Essen, Germany
Fritz Schupp and Martin Kremmer, 1932/1961

Aerial view of the Zeche Zollverein coal mine industrial complex in Essen, Germany, ca. 1934.

The first coal mine shaft near Essen was opened in 1847 and was named Zollverein after the German Customs Union, Deutscher Zollverein, which had been founded in 1834. By 1914, the Katernberg area near Essen, in the state of North Rhine-Westphalia (NRW), had become home to four independent mines, with a total of eleven shafts. In order to ensure its economic survival after WWI, the plant systems were consolidated and mechanized for efficiency and high performance with largely automated work processes.[1] Zollverein Shaft XII was designed for this purpose in 1927 by German architects Fritz Schupp (1896–1974) and Martin Kremmer (1894–1945) and completed in 1932. Together with the corresponding above-ground structures, it was designed to improve the system of coal extraction and processing. Undoubtedly inspired by the Bauhaus, which had moved in 1926 to Gropius' building in Dessau-Roßlau, Schupp and Kremmer's design for the industrial buildings embodied the school's shift towards Cubism and Functionalism. Their double-truss pit frame for Zeche Zollverein has become an icon of the region's industrial culture.[2] The architects created a symmetrical and geometrical complex arranged along two axes, which resulted in a unique industrial plant model, often considered to be the "most beautiful mine in the world."[3] Fritz Schupp used the same architectural vocabulary for

Exterior view of the double-truss pit frame at Shaft XII, 1986.

the coking plant which was completed in 1961. Zeche Zollverein was the largest and most powerful coal processing plant in the world after 1932 and right up until the mine closure in 1986. The coking plant remained open until 1993.

In 2001, the complex was included in the UNESCO list of World Heritage Sites, cited as being "a monument of industrial history reflecting an era in which, for the first time, globalization and the worldwide interdependence of economic factors played a vital part."[4] The World Heritage designation also states that Zeche Zollverein's buildings are "outstanding examples of the application of the design concepts of the Modern Movement."[5] Exceptional value was attributed to the Bauhaus-influenced architecture of the industrial complex, which for decades provided the model for modern industrial construction.[6] Today it is one of the anchor points of the European Route of Industrial Heritage and the whole industrial complex is an ideal place to learn about the mining history and development of industrial architecture in one of Europe's most significant industrial regions.[7]

CONDITION PRIOR TO INTERVENTION

The exterior architecture of Zeche Zollverein is characterized by large expanses of red brick infill laid within the orthogonal grid of its exposed post-and-lintel steel framework (which is painted red). Set within the masonry infill are wide bands of fixed and operable steel-framed ribbon windows that are single-glazed with 1/4" (6mm) thick cast glass to provide diffused daylight to the interior. A structural skeleton system is used throughout most of the complex, consisting of full building-height steel portal frames (that remain exposed) and a concrete-encased steel frame construction forming the floor and roof decks.

The thin exterior single-wythe brick wall construction and the non-thermally broken single-glazed steel frame windows have poor thermal performance, making it more difficult to adapt the facades to other uses. Fortunately, the buildings were well-maintained during the years between 1986 and 1993, and there was no significant damage before the interventions began. It also helped that some of the buildings in the complex were used for temporary cultural events starting as early as 1987, and so had been kept in good condition.

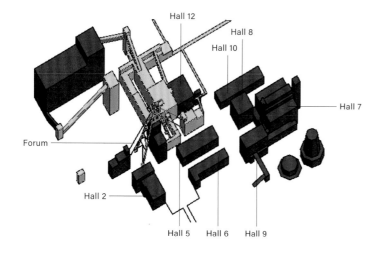

Axonometric overview of Shaft XII.

Rear (east) view of the buildings at Shaft XII, 2013.

INTERVENTION

Most of the buildings in the complex were refurbished (primarily by architects Heinrich Böll and Hans Krabel) and reused for permanent cultural and historical purposes in 1989.[8] Each building was treated differently, which has made the complex a showcase of facade refurbishment at its best. Interventions have included: simply restoring the steel profiles with the existing single-glazed panes or with new wired glass panes (Halls 2 and 5); adding new slim IGUs partly with wired glass panes (Hall 7); installing a thermally broken frame with IGUs with wired glass (Hall 9); and installing insulated partitions with secondary glazing consisting of single-pane steel frame casement windows set within an interior insulated partition built parallel to the existing exterior walls, thereby creating a large insulating air cavity (Halls 6 and 10).

Since 1992, the former central workshop (Hall 5) has been used as an exhibition space that required only moderate heating up to 14°C maximum. This circumstance maintains the building's austere industrial character internally and allowed the steel frame windows to be restored with their original single glazing. In non-glazed areas an additional interior layer of lightweight perforated bricks with a lime cement mortar finish was added to the inboard side of the exterior walls to improve thermal performance. The hall famously hosted the work of Ulrich Rückriem as part of the *Documenta 9* art exhibition in 1992. Between 1996 and 1997, the Boiler House (Hall 7) and the former Low Pressure Hall (Hall 9) were transformed into the Red Dot Design Museum and the restaurant

Steel frame windows at the former Boiler House (Hall 7), current site of the Red Dot Design Museum, 2007.

When there were lower thermal performance requirements (e.g. in Halls 2 and 5), the original steel frame single-glazed windows were restored or were retrofit with new single-pane wired glass, 2005.

Broken single-pane wired glass, 2014.

Casino Zollverein. Both adaptations demanded higher comfort requirements. At Hall 7 (museum), the approach to the exterior was to restore the building facade and remove several later additions in order to reveal its original form, while in the interior "the heavy industrial feel of the building was maintained."[9] The original single glazing was replaced with new IGUs within the existing steel frames (installed from the interior to prevent the costly use of scaffolding). For economic reasons, the interior is only moderately heated to a maximum of 16°C. Altogether, the interventions maintain the calm and homogeneous industrial appearance both inside and outside.

Where higher thermal performance was required (e.g. in Halls 7 and 9), the original steel frames were restored and rehabilitated by adding new slim IGUs, 2005.

There were higher requirements for indoor climate and thermal comfort for the restaurant in Hall 9. To address this, a replacement thermally broken cold-formed steel frame window system with 1" (25mm) thick IGUs was developed using the same sightlines as the existing windows. The new windows were installed in a new interior stud wall that was built parallel to the existing exterior masonry wall. The new windows align with the original single-glazed steel frame units and provide them with secondary glazing to deliver a high level of thermal protection while maintaining the industrial interior appearance. Some of the original single-glazed wired glass in Hall 9 was replaced with transparent IGUs in an effort to enhance a visual connection to the exterior.

The same intervention was implemented at the former electrical workshop (Hall 6) when it was converted into an exhibition hall and store. To allow for the additional interior stud wall and

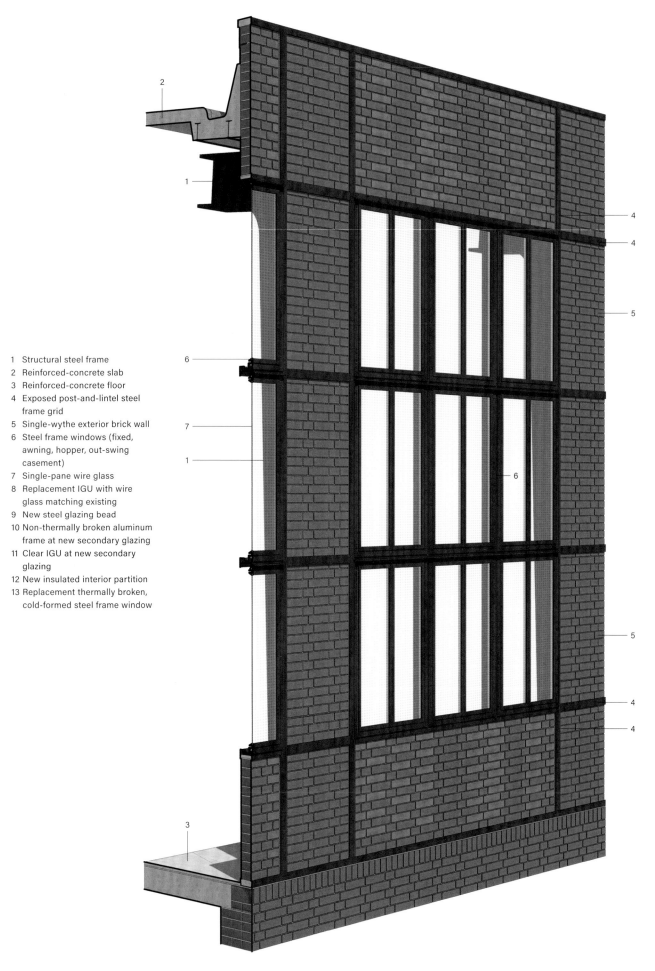

1 Structural steel frame
2 Reinforced-concrete slab
3 Reinforced-concrete floor
4 Exposed post-and-lintel steel frame grid
5 Single-wythe exterior brick wall
6 Steel frame windows (fixed, awning, hopper, out-swing casement)
7 Single-pane wire glass
8 Replacement IGU with wire glass matching existing
9 New steel glazing bead
10 Non-thermally broken aluminum frame at new secondary glazing
11 Clear IGU at new secondary glazing
12 New insulated interior partition
13 Replacement thermally broken, cold-formed steel frame window

Typical section at original steel frame window.

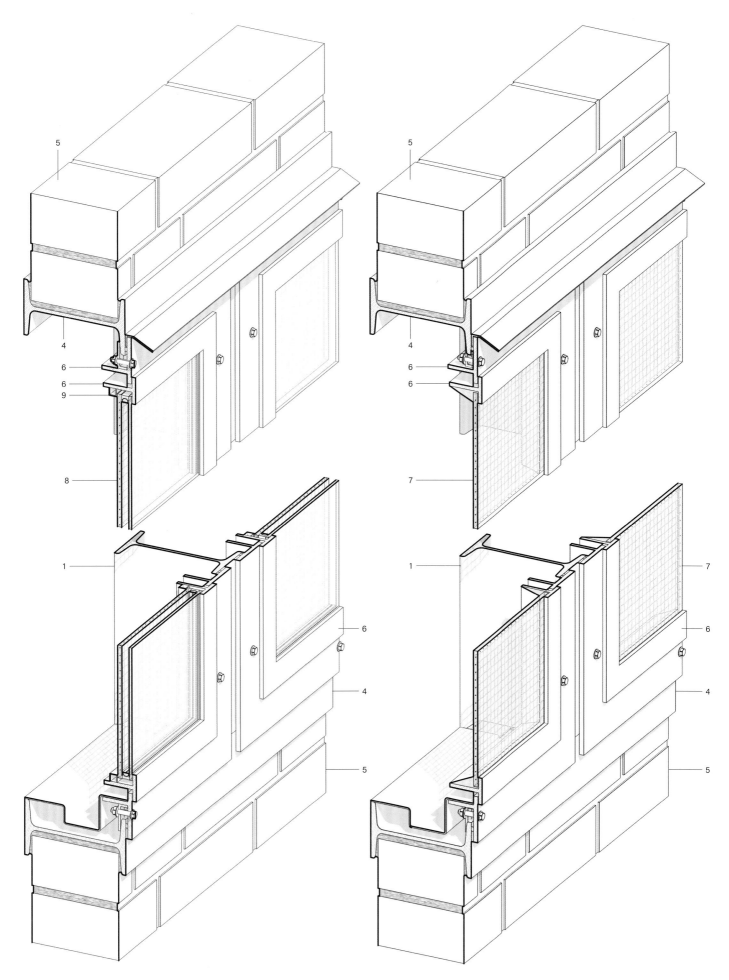

Typical detail at restored steel frame window and new secondary glazing.

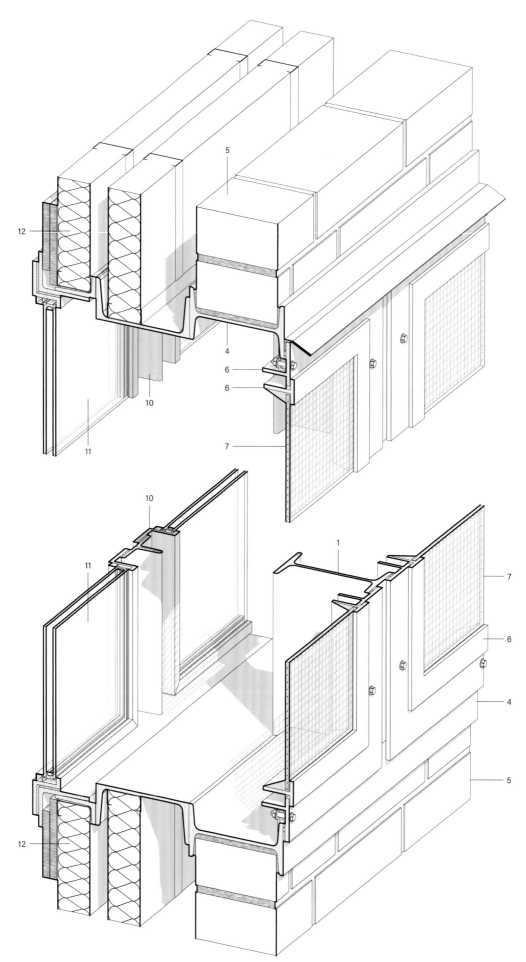

Typical detail at rehabilitated steel frame window with IGU.

1 Structural steel frame
2 Reinforced-concrete slab
3 Reinforced-concrete floor
4 Exposed post-and-lintel steel
  frame grid
5 Single-wythe exterior brick wall
6 Steel frame windows (fixed, awning,
  hopper, out-swing casement)
7 Single-pane wire glass
8 Replacement IGU with wire glass
  matching existing
9 New steel glazing bead
10 Non-thermally broken aluminum
  frame at new secondary glazing
11 Clear IGU at new secondary glazing
12 New insulated interior partition
13 Replacement thermally broken,
  cold-formed steel frame window

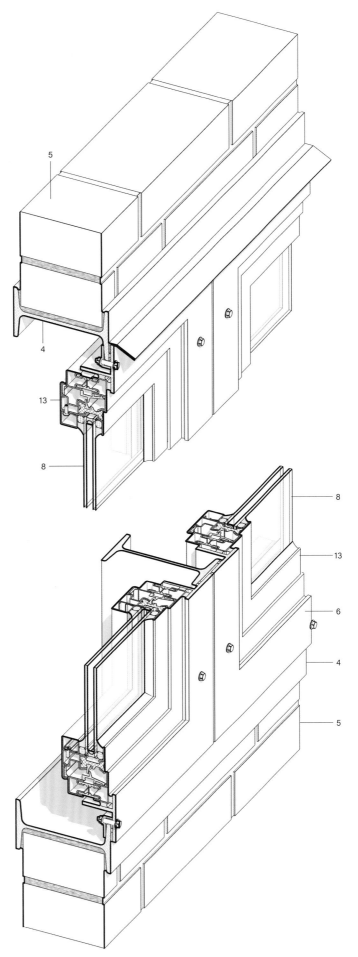

Typical detail at thermally broken cold-formed steel frame window with IGU.

Example of the restored original steel frames fitted with new slim IGUs that were installed in Halls 7 and 9, 2005.

Example of the restored original steel frames fitted with new slim IGUs that were installed in Halls 7 and 9, 2005.

Exterior view of restored steel frame single-glazed awning windows and adjacent secondary glazing installed to meet the higher thermal performance requirements of interior spaces that are in continuous use, such as in Halls 6 and 10, 2005.

secondary glazing, the inner pumice-concrete block facing was removed and supplemental rails were installed to anchor and secure the historic facade. Overall, the steel post-and-lintel framing retained its structural function and refurbishment was not necessary for the most part. The exception was at the perimeters of the ribbon windows, gates and door openings of the existing post-and-lintel structure which were refurbished "by applying a complete rustproof coating or by replacing individual members." All other facade steel profiles were sandblasted and coated on their exposed surfaces only.[10]

A different approach of additional interior layers was chosen for the former warehouse (Hall 10). There, a box-within-a-box concept was utilized to accommodate permanent spaces such as workshops, stores, classrooms, offices, changing rooms and sanitary facilities. By constructing a new thermally insulated structure within the existing hall, it was possible to leave the historic facade intact without any modifications. The corridors between both structures served as a thermal buffer and are used seasonally for teaching and administrative purposes.

The coal mine complex and the coking plant have been owned by the Zollverein Foundation since 2009 and 2010 respec-

Front (west) view of the building at Shaft XII, 2013.

tively. New buildings have been added to the adapted historical buildings at the complex, extending its new cultural functions. Designed by the Japanese architectural office Sanaa, the 'Zollverein Cube' was built in 2006, the first new building at Zollverein in 50 years. Its thin rectangular facades, cubic shape and the oversized room heights are reminiscent of the iconic designs of Schupp and Kremmer. The Zollverein Cube was used by different design schools until 2017, and today is part of the Folkwang Zollverein World Heritage Campus, serving as a starting point for a future "design city" aimed at building up a larger design campus. In 2017, MGF Architects designed a new campus in the Northern Quarter to host the design department of the Folkwang University of the Arts. The new building's orthogonal facades made of flush grey galvanized steel plates interpret Zeche Zollverein's industrial character qualitatively rather than imitating it.

COMMENTS

The variety of approaches to the rehabilitation of the steel frame glazed enclosures at Zeche Zollverein is unique within the field of conservation of Modern architecture. The number of different interventions is directly related to the diversity of uses and new

View of the interstitial space between the restored steel frame single-glazed windows (left) and adjacent secondary glazing (right) that was installed in Halls 6 and 10, 2005.

The Zollverein Cube, designed in 2006, was the first new building constructed at the Zollverein complex in fifty years, 2009.

Restored steel frame in-swing hopper window, 2008.

functions added to the complex over time, with each program requirement resulting in a unique tailored approach. As the industrial complex was still in use and not modified prior to the first intervention, the original steel frame glazed enclosures were mostly intact. For this reason, Zeche Zollverein's glazed enclosures are essentially an authentic as-built documentation of its original design—characterized by its large-scale uniform detailing and material usage.

The challenge for the rehabilitation and conservation of the complex was not in the identification of the original members and materials, but rather in the decision-making process around adapting the "simple" industrial envelopes to the new uses and functions, each with different requirements and standards that had to be taken into consideration. Throughout the process, it was critical that the complex's recognizable exterior appearance, with its unique and elegant industrial materiality, be retained. To ac-

CREDITS

Zeche Zollverein
(1932/1961)
Essen, Germany

ORIGINAL CONSTRUCTION

Architects
Fritz Schupp and Martin Kremmer

1989–2019 INTERVENTIONS

Client
Bauhütte Zeche Zollverein
Schacht XII
GmbH/Stiftung Zollverein

Coordinating/Contact Architects
Böll and Krabel Architekten
Heinrich Böll Architekt, Essen

Other Architects
Foster and Partners, London
OMA/Rem Koolhaas,
Rotterdam

SANAA, Kazuyo Sejima + Ryue Nishizawa, Tokyo
MGF Architekten GmbH,
Stuttgart

When there were lower thermal performance requirements (e.g. in Halls 2 and 5), the original steel profiles and existing single-glazed panes were restored or retrofit with new single-pane wired glass, 2005.

complish this, each building and its interventions were treated differently, particularly when it came to their specific thermal requirements, where a range of interventions to the glazed enclosures were applied. As part of the conservation plan devised for the complex, the interior spaces were also left as much in their original state as possible, often with remaining equipment in place, in order to retain and highlight the authentic industrial atmosphere. Across the entire complex only limited architectural interventions were made to adapt to the new functions, adding elements such as new staircases and elevators (like the large external escalator leading to the Ruhr Museum at the coal washing plant, for example) and sometimes creating new levels to provide access into the former industrial areas and to specific functions, such as new kitchens and offices.

The diverse solutions implemented throughout the complex range from subtle restorations that are hardly noticeable to new additions that are clearly readable (such as the new interior walls and secondary glazing which are visible from the exterior due to their silver aluminum finishes). These localized changes in the historic appearance do not distract from experiencing the complex's unity as an architectural ensemble though. The long-term effort of Heinrich Böll and Hans Krabel as the architects-in-charge is certainly commendable—under their guidance, selective and well-purposed modifications and new additions have been undertaken. Many suitable approaches were created and implemented with a high degree of technical craft and skill, which has been rewarded with several architectural accolades.

NOTES

1 Zollverein, "What happened until now," accessed 30 December 2018, https://www. zollverein.de/ueber-zollverein/geschichte/.

2 Frieder Bluhm, "Symbol für Zuversicht und Wandel. Unesco-Welterbe Zollverein in Essen," *Industriekultur* (March 2013): p. 31.

3 World Heritage Germany, "The 'most beautiful coal mine in the world': the Zollverein UNESCO World Heritage Site Tourist development, products and highlights," last modified 21 January 2018, accessed 3 February 2019, https://worldheritagegermany.com/the-most-beautiful-coal-mine-in-the-world-the-zollverein-unesco-world-heritage-site-tourist-development-products-and-highlights/.

4 UNESCO, "World Heritage List - Zollverein Coal Mine Industrial Complex in Essen," accessed 4 November 2018, http://whc.unesco.org/en/list/975.

5 Ibid.

6 UNESCO. "UNESCO-Welterbe Industriekomplex Zeche Zollverein in Essen: Industriedenkmal im Stil des Bauhauses," accessed 31 December 2018, https://www.unesco.de/kultur-und-natur/welterbe/welterbe-deutschland/industriekomplex-zeche-zollverein-essen.

7 For further information, see www.zollverein.de (home page of the Zollverein complex) and www.erih.net (home page of the European Route of Industrial Heritage).

8 Heinrich Böll and Hans Krabel, *Arbeiten an Zollverein. Projekte auf der Zeche Zollverein Schacht XII seit 1989* (Essen: Klartext-Verlag, 2010), p. 147.

9 Foster and Partners, "Red Dot Design Museum," accessed 7 December 2018, https://www.fosterandpartners.com/projects/red-dot-design-museum/.

10 Heinrich Böll and Hans Krabel, "Zollverein Coal Mine, Pit XII in Essen – Conversion and Extension of an Industrial Monument Dating from 1928–1932," *Detail* 6 (1997): p. 877.

Van Nelle Factory

Rotterdam, Netherlands
Brinkman & Van der Vlugt, 1931

Northeast facade of the Van Nelle Factory in Rotterdam, Netherlands, before the installation of the overhead transport bridges with conveyor belts, ca. 1931.

Completed in 1931, the Van Nelle Factory in Rotterdam imported tobacco, tea, coffee and fabrics and offered a roasting house, storage and support facilities, all linked by diagonal overhead bridges that established dynamic and seamless connections between supply, manufacturing and distribution lines. The complex was designed by Dutch architect Leendert van der Vlugt (1894–1936) from the local architectural office of Brinkman & Van der Vlugt. Dutch architect and civil engineer Jan Gerko Wiebenga (1886–1974), a pioneer of reinforced concrete design in the Netherlands, was a key collaborator in the conception of this large storage and manufacturing facility located along the banks of the Delfshavense Schie Canal.

The complex's simple and efficient design relies on the extensive use of a translucent steel frame curtain wall system consisting of flat 4" (10cm) deep non-thermally broken steel mullions located every 3'–3 3/8" (1m) and spanning 12'–1 3/8" (3.7m) vertically floor-to-floor. The typical curtain wall module between mullions consists of three panels of the same height. The bottom panel is an insulated metal spandrel. The upper panel is made of fixed single-pane units with 1/4" (6mm) thick glass. The middle panels alternate between fixed units identical to the top panels and verti-

Curtain wall with missing and replacement spandrel panels and other alterations, 1980s.

Restored steel frame single-glazed curtain wall, 2016.

cal pivot windows made of standard 1 3/8″ (35mm) thick steel windows manufactured by Crittall.[1] The curtain wall is suspended from the edge of flat reinforced-concrete slabs, which are supported by octagonal columns with mushroom heads. The spandrels are made of two steel sheets with a 1 3/16″ (30mm) thick panel made of Torfoleum (a material comprising impregnated peat used at the beginning of the 20th century for insulation purposes). The exterior side of the spandrel panels and the window sash were sprayed with a zinc-rich coating. The glass panes were made of drawn glass, a manufacturing process that predates contemporary float glass and has been mostly discontinued. The windows could be cleaned from an exterior cage suspended from an overhead rail.[2] The combination of high ceilings, glazed enclosures, daylit interiors and a large-span concrete structure resulted in flexible and functional open plans that embraced early American and European concepts of Modern efficient factory design. After visiting the Van Nelle Factory shortly after it opened, Le Corbusier praised it as "the most beautiful vision of modern times" known to him and went on to state that "if the current world was similarly organized, harmony would be the crowning achievement of our labor."[3]

In 2014, the complex was listed by UNESCO as a World Heritage site. The World Heritage designation states that the Van Nelle Factory "embodies the new kind of factory that became a symbol of the modernist and functionalist culture of the inter-war period and bears witness to the long commercial and industrial history of the Netherlands in the field of importation and processing of food products from tropical countries, and their industrial processing for marketing in Europe."[4]

CONDITION PRIOR TO INTERVENTION

By 1995, when all production was stopped in the Van Nelle Factory, the complex had been in continuous use for more than 60

TIMELINE

1931	Completed
1985	Registered as a Dutch National Monument (Rijksmonument)
1995	Production ceased
1998	National and municipal Departments for Conservation of Architectural Heritage acquired the complex to convert it into the Van Nelle Design Factory
1999	Wessel de Jonge Architects develops master plan
1999–2004	Redevelopment completed
2014	Listed as a UNESCO World Heritage Site

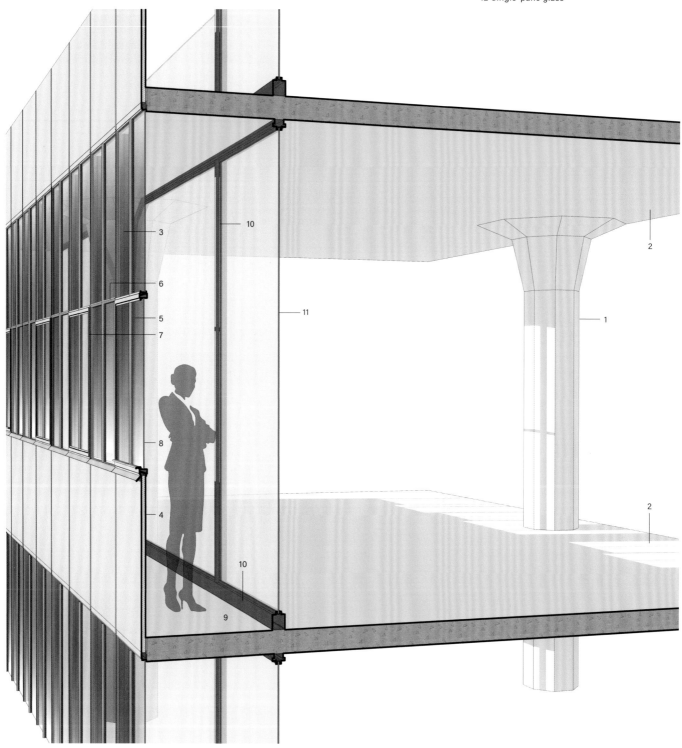

1 Reinforced-concrete column
2 Reinforced-concrete slab
3 Steel curtain wall mullion
4 Insulated steel curtain wall spandrel
5 Steel curtain wall muntin
6 Steel curtain wall transom
7 Steel curtain wall window
8 Single-pane clear glass
9 New circulation space
10 New secondary glazing aluminum frame
11 Single-pane laminated glass
12 Single-pane glass

Typical section at existing rear (southeast) facade steel frame curtain wall and new secondary glazing after rehabilitation.

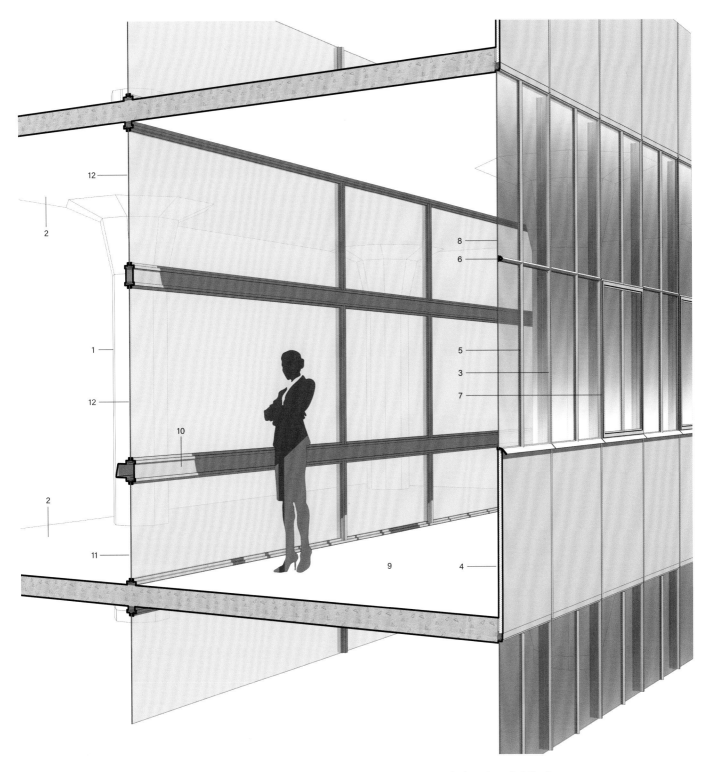

Typical section at existing front (northeast) facade steel frame curtain wall and new secondary glazing after rehabilitation.

Interior view of the restored original steel frame single-glazed curtain wall, 2016.

Interior view of the original steel frame single-glazed curtain wall at the former coffee factory building, 2016.

years. Several buildings had been added to the complex and there were signs of wear and tear throughout. Original metal spandrel panels were missing at selected curtain wall locations and had been replaced with other materials. Glass panes at both fixed and operable units had been replaced as needed or altered to allow for the installation of exhaust vents. The metal spandrel panels and steel frames exhibited signs of corrosion. With all the machinery removed from the interiors and the buildings empty, the vitality and dynamism of the complex's heyday was gone.

INTERVENTION

The intervention approach devised by the team to guide the renovation and reuse of the Van Nelle complex relied on the notion that the "material authenticity" of character-defining features, such as the mushroom column capitals and the steel facade panels, were secondary to their "design authenticity."[5] In other words, the degrees to which the original design concepts could still be perceived was deemed more relevant than the degree to which their materials and constructions are authentic. For the steel frame curtain wall, that meant reinstating its physical integrity, as well as the transparency that ensured bi-directional views from and towards the building interiors. To achieve this, broken glass was replaced with new drawn glass units produced in the Czech Republic, which matched the characteristics of the original glass. The operable vertically pivoted windows were made operational and repainted with a metallic coating. Missing original entrances at the ground floor were restored.[6]

Secondary glazing (right) at the new circulation space adjacent to the restored original steel frame single-glazed curtain wall (left), 2016.

Along with the restoration of the original steel frame curtain wall, the adjacent mechanical systems were refurbished. As the exterior wall assembly was initially conceived for manufacturing use and storage and would not meet today's performance requirements for offices, additional modifications to the building's mechanical systems were required to accomplish the building's adaptive reuse. To address concerns about the poor performance of the original exterior wall assembly, additional heating was provided through a new radiant system and a new high-performance floor-to-ceiling thermally broken aluminum frame glazed enclosure (manufactured by Oskomera) was installed parallel to the original curtain walls on the northeast facade. The resulting approximately 8' (2.5m) wide transitional space between the low-performing original enclosure and its adjacent addition is used for circulation. This intermediate zone also acts as a buffer protecting the new interior spaces from both the exterior climate and noise from adjacent train tracks and highways. On the southwest facade, the space between the original restored facade and the new secondary glazing is only about 2'–6" (0.8m) wide—enough to be used as an active double-glazed envelope where the interior temperatures are controlled year-round through a

Secondary glazing and new circulation space adjacent to the restored original steel frame single-glazed curtain wall, 2016.

Restored original steel frame single-glazed curtain wall and new secondary glazing, 2016.

CREDITS

Van Nelle Factory (1931)
Rotterdam, Netherlands

ORIGINAL CONSTRUCTION

Architects
Leendert van der Vlugt
(Brinkman & Van der Vlugt)
Jan Gerko Wiebeng
(Civil Engineer)

1999–2004 INTERVENTIONS

Client
CV Van Nelle Ontwerpfabriek

Development/Management
Maatschap Van Nelle
Ontwerpfabriek

Coordinating Architects
Wessel de Jonge Architecten
BNA BV

Historic Building Survey
Suzanne Fischer and Wessel
de Jonge Architecten BNA BC

Architectural Paint Research
Polman Kleur & Architectuur

Structural Engineers
ABT building technology
advisers

Architects for the Factory
Claessens Redmann Architects
& Designers BV

**Architects for Dispatch
Building**
Wessel de Jonge Architecten
BNA BV

Architects for Office Building
Molenaar & Van Winden
Architecten

Landscape Architects
DS Landschapsarchitecten

Building Physics Consultant
Climatic Design Consult

**Structural Engineer
Consultant**
Ingenieursbureau Bartels

General Contractor
IBB Kondor BV

Security
PreNed Beveiligingstechniek

combination of mechanical services, window treatments (blinds) for shading and glare control, and natural ventilation through the operable windows.[7]

COMMENTS

The adaptive reuse of the Van Nelle Factory is a successful intervention that balances the authenticity of the original design with the requirements of the new uses. The new secondary glazing installed in the former factory spaces is designed so that it can be clearly differentiated from the original adjacent curtain wall. The height of the new interior spandrel panels, the width and configuration of the new horizontal vision glass, the make-up of the low-reflection glass panes, the industrial design of the sliding doors, as well as the low height of the interior partitions with all-glass clerestories above to expose the mushroom column capitals, and even the lighting placement, fixture selection and color temperature are all conceived to enhance the original design concept, as well as daylighting and views through the original enclosure.

The design of the mechanical systems serving the spaces within the new secondary glazing included fresh air intakes routed through acoustically conditioned grilles at the bottom of the new glazed enclosures. This design feature draws in exterior air (from the shaded northeast in the summer and from the sunlit southwest in the winter) into the interior spaces by taking advantage of the draftiness of the original curtain wall, which has no weatherstripping. This simple, environmentally friendly approach turns one of the less desirable features of the original curtain wall assembly into an opportunity for integrated design.[8] The sensitive box-within-a-box solution implemented in the Van Nelle Factory enhances the authenticity of the original design without compromising the material integrity of the steel frame curtain wall and the reinforced-concrete structure. It affords the large complex a new use for its post-industrial life, and does so while providing the building with new environmental features and a dynamic circulation system that enhances a sense of place rooted in the site's industrial past.

NOTES

1 David Yeomans, "The Pre-history of the Curtain Wall," *Construction History* 14 (1998): p. 60, accessed 27 January 2019, http://www.jstor.org/stable/41601861.

2 "The Van Nelle Factory, Rotterdam: J.A. Brinkman and L.C. Van der Vlug, Architects," *Architectural Record* 69 (May 1931): p. 417.

3 Le Corbusier, "Voyage D'hiver…Hollande," *Plans* 12 (February 1932): p. 40.

4 UNESCO, "World Heritage List – Van Nellefabriek – Description," accessed 4 June 2016, http://whc.unesco.org/en/list/1441/.

5 Wessel de Jonge, "Continuity and Change in the Architecture of Van Nelle," In *Van Nelle: Monument van de vooruitgang*, ed. J. Molenaar, p. 260 (Rotterdam: De Hef, 2005). Accessed 27 January 2019, http://www.wesseldejonge.nl/media/downloads/van_Nellefabriek_ENG_compr.pdf.

6 Ibid., pp. 263, 283.

7 Ibid., p. 263.

8 Ibid., p. 266.

Hardenberg House

Berlin, Germany
Paul Schwebes, 1956

Hardenberg House in Charlottenburg, Berlin, Germany, 1957.

Hardenberg House, one of the most significant commercial buildings of post-war Modernism in West Berlin, was designed by German architect Paul Schwebes (1902–1978).[1] Built between 1955 and 1956, it was from the beginning the headquarters of Kiepert, a long-established independent bookstore and second-hand bookshop in Berlin. In his design, Schwebes took the signature elements of pre-war Modernism—including the exposed structural elements, window configurations and the use of colors and materials—and further developed them. He seemed to have been inspired by the designs of Erich Mendelsohn, as evidenced in the dynamic sweep of the three front facades with their characteristic horizontality.[2]

The seven-story reinforced-concrete structure follows the configuration of the adjacent streets with an elegant, curved plan layout that affords the building a continuous facade plane stretched across three streets and three more toward the courtyard. The front (street) facades are divided into three vertical zones reminiscent of a Classical design composition. The ground-floor retail storefronts are made of brass-colored anodized aluminum frames. They are slightly set back from the street facades and sheltered by a continuous awning.[3] The five office floors above are

Front facades after the rehabilitation work, 2013.

delineated by steel frame window walls, and a large slim cantilevered eave emphasizes the roof penthouse. The floor slab edges prominently separating the window walls at the office floors are clad with white Detopakglas (an opaque glass with a coated surface facing the interior of the building). The same opaque glass material in black is also used for the spandrel panels. The window walls are comprised of non-thermally broken steel mullions and window units, each with a large fixed vision panel in the center flanked by slender in-swing casements. The mullions supporting the windows and spandrels are made of narrow steel profiles with a brass-colored finish. All of the windows have slender steel frames, which are 2″ (49mm) wide and 3″ (75mm) deep. The in-swing steel frame casements, which are painted black on the exterior and white on the interior, have supplemental in-swing interior glazed shutters painted the same white color as the main casements. This window arrangement of double casements forms an air cavity when the sashes are together, which improves thermal insulation and allows for maintenance when required. The slender window profiles and the white and black Detopakglas at the slab edges and spandrels, together with the brass-colored mullions, lend a distinctive, elegant appearance to the building street facades. Exterior roll-down white sunscreens were included in the facade design to provide sun protection during the summer.

While the front facades are composed primarily of large glazed window walls, the rear (courtyard) facades are made of white-plastered masonry and punched steel windows similar to those on the front facades. Semi-circular steel frame glazed staircases with translucent opaque glass project off the rear facade planes conveying a unique sense of dynamism and rhythm.

TIMELINE

1955	Construction started
1956	Construction completed and building opened as headquarters of bookstore Kiepert
1973–1974	Adaptation of the interior by Günther Wehde
2001	Removal of neon signs from shop windows
2002	Insolvency of the Kiepert bookstore; Hühne Immobilien buys the building
2003	Building reopens; Lehmann (a bookstore) occupies three floors
2003–2004	Winkens Architekten oversees first and only renovation; building is listed
2004	Rehabilitation project is awarded the Federal Prize for Craft in Conservation
2018–2019	Building hosts the temporary Bauhaus-Archiv and Bauhaus-Shop while the original Bauhaus-Archiv in Berlin is under renovation

Restored concrete structure and replacement aluminum frame glazed enclosure at one of the rear facade staircases, 2019.

CONDITION PRIOR TO INTERVENTION

Kiepert remained the main building tenant until their insolvency in 2002, after which the building was sold. During the 50 years of their occupancy, little had been done to the building in the way of maintenance and repair. There were broken glass panes and the steel profiles were substantially damaged by corrosion. The glazing of the staircases at the rear facades was also severely damaged due to previous inappropriate repairs.[4]

The window-to-wall ratio at Hardenberg House (generally over 50%) was the main weakness of its 1950s-era glazed front facades due to the high thermal loads it caused in the summer, particularly after the sunscreens were damaged. On the other hand, when compared to the typical depth of the floor plates of other buildings from the 1950s, the extensive glazing at the street facades allowed for good day lighting and, with two operable casements per window bay, natural ventilation. Other issues of concern, such as leaks related to steel frame corrosion and lack of maintenance to the double casement windows (which eventually compromised their thermal performance), could not be addressed without modifying the overall building configuration and design.

INTERVENTION

When the real estate company Hühne Immobilien took over as the new building owner, they found out that the current energy conservation requirements and window U-values could not be achieved with the existing facade elements. They decided to make the necessary window replacements at the rear facade and to

Corrosion of the steel frame glazed enclosures and damaged interior concrete finish at one of the rear facade staircases, 2003.

Restored concrete structure and stairs with replacement aluminum frame glazed enclosure at one of the rear facade staircases, 2019.

Restored steel frame window wall with brass-colored mullions, black glass spandrel panels and white glass slab edge covers at one of the street facades, 2005.

Interior view of the restored steel frame window wall with secondary glazing at one of the street facades, 2007.

keep the original appearance of the front facades as much as possible. Thus, the rehabilitation aimed to maintain the spirit of Hardenberg House' 1950s architecture and to preserve the building's elements and materiality as much as possible.

The focus of the rehabilitation was to restore the front facades and so the original construction was repaired carefully. The original steel window frames were dismantled and the remaining finish coatings and rust were removed, before the same frames were repainted and reused. The original seals were replaced, and where needed (mostly due to damage) the glass panes were replaced with new single glazing of the same 1/4" (6mm) thickness. The 1/4" (6mm) thick glass at the inner sash of the double casements was replaced with a 1/4" (6mm) thick Pilkington K Glass™, which has a low-emissivity coating, on glass surface #3 (towards the cavity) and improves the overall U-value of the assembly by reflecting heat back into the room while still allowing passive solar gain.[5] New glazing beads where provided for both casements, and concealed weather-stripping was installed on the exterior sash. The corroded brass-colored mullions were replaced in-kind and window hinges and other hardware were retrofitted. The original white sun protection blinds were also replaced with new brass-colored ones.[6]

The rear facades were rehabilitated with replacement windows. The original glazing of the stairwells was also replaced with a new thermally broken and insulated aluminum and glass enclosure. The two side staircases, previously used as secondary entrances, received the most extensive intervention when they were altered to provide direct access to the street and upgraded with a design as prestigious as the main entrance. The rehabilitation work also included a lighting design that incorporated the canopy

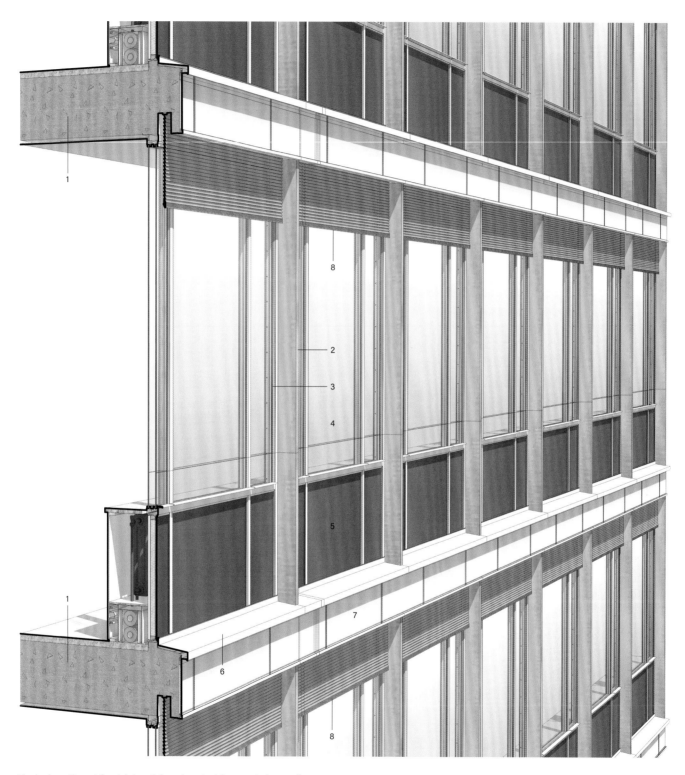

Typical section at front (street) facades steel frame window wall.

1 Reinforced-concrete slab
2 Steel window wall mullion with
  brass-colored finish
3 Steel frame in-swing casement window
  and existing secondary glazing
4 Steel frame fixed window
5 Black-coated glass window wall spandrel
6 White-coated metal slab edge cover
7 White-coated glass slab edge cover
8 Exterior blinds
9 Single-pane clear glass
10 Replacement single-pane laminated glass
   with low-e coating at secondary glazing
11 Replacement steel glazing bead
12 Replacement weatherstripping

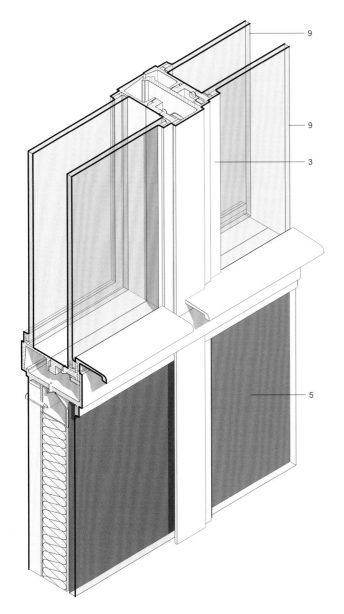
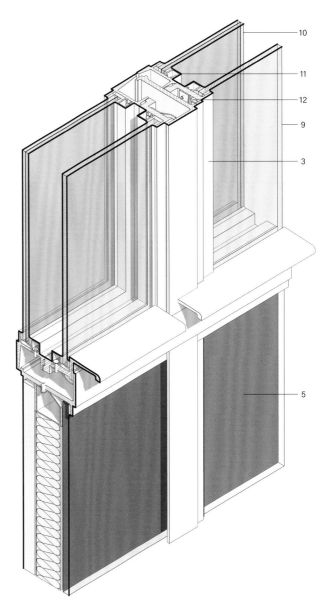

Typical before (left) and after (right) window wall details.

View of the restored steel frame window wall looking up to the cantilevered eave at a corner of the street facades, 2005.

and installed an advertising ribbon over the ground-floor shop windows. In response to comfort and security requirements, the building services were also partly modernized, providing heating and ventilation to the shops and offices, as well as air conditioning.[7]

COMMENTS

An important aspect of this renovation was that the front and rear facades were treated separately. This approach is consistent with the original design intent which emphasized the street fronts as the main design components and reduced the role of the rear courtyard facades to simply functional. By optimizing U-values and avoiding thermal loss at the rear facades, the owner was able to improve the overall energy performance of the building while

CREDITS

Hardenberg House (1956)
Berlin, Germany

ORIGINAL CONSTRUCTION

Architect
Paul Schwebes

1993–2006 INTERVENTIONS

Client
Hühne Immobilien

Architects
Winkens Architekten, Berlin
(LP 2–5)

Tender, Site Management
Ingenieurbüro Welke +
Schönepauck, Berlin
(LP 2–5)

Facade Construction
Fensterfabrik Montag,
Biberach-Mettenberg

Detail of brass-colored anodized aluminum frames at the ground floor storefront, new infill panels and white slab edge covers above at one of the street facades, 2008.

retaining the authenticity, materiality and original appearance of the front facades. Thus, the compromise (of replacing original material with new insulated window and wall installations at the rear facades) arguably allowed for a higher level of conservation to be achieved at the front facades.

Better energy performance was achieved from the front facades with only minimal modifications to the original aesthetic of the double window construction. This was largely possible due to the skill exercised in adapting the steel frame window elements to meet contemporary requirements of thermal protection. Workers needed to be specially trained in dismantling, repairing and reinstalling the existing assemblies, which often had to be individually adapted to specific tolerances or to adjacent building elements. Another change seems to have been simply fortuitous—replacing the original white blinds with new brass-colored ones was probably an aesthetic decision, but it also served to reduce the amount of sun reflection coming into the interior, resulting in less glare and overheating.

The commitment of the rehabilitation team in working with the owner and the preservation authorities has resulted in a very successful rehabilitation for the Hardenberg House, and has earned them several awards, among them the Bundespreis für Handwerk in der Denkmalpflege (Federal Prize for Craft in Conservation).[8]

NOTES

1 BerlinOnline Stadportal, "Über den Bezirk – Geschaeftshaeuser," accessed 9 December 2012, https://www.berlin.de/ba-charlotten-burg-wilmersdorf/ueber-den-bezirk/gebaeude-und-anlagen/geschaeftshaeuser/artikel.158744.php.

2 Mathias Remmele, "In die Jahre gekommen: Haus Hardenberg," *Deutsche Bauzeitung* 144, no. 11 (2010): p. 53, accessed 29 December 2018, https://www.db-bauzeitung.de/db-themen/db-archiv/haus-hardenberg/.

3 Uta Pottgiesser, "Revitalisation strategies for modern glass facades of the 20th century," *Structural Studies, Repair and Maintenance of Heritage Architecture XI, WIT Transactions on The Built Environment*, vol. 109, ed. C.A. Brebbia (Wessex Institute of Technology, WIT Press, 2009), p. 572.

4 Remmele, "In die Jahre gekommen," p. 55.

5 Pilkington, "Pilkington K Glass," accessed 29 December 2018, https://www.pilkington.com/en-gb/uk/products/product-categories/thermal-insulation/pilkington-k-glass-range/pilkington-k-glass.

6 Pottgiesser, "Revitalisation strategies," p. 572; Dirk Dorsemagen, "Büro- und Geschäftshaus-fassaden der 50er Jahre: Konservatorische Probleme am Beispiel West-Berlin," (doctoral thesis, Technische Universität Berlin, 2004), pp. 314–316, accessed 29 December 2018, http://dx.doi.org/10.14279/depositonce-945.

7 Remmele, "In die Jahre gekommen," p. 55.

8 BerlinOnline Stadportal, "Bundespreis Handwerk – Preisverleihung 2004," accessed 29 December 2018, https://www.berlin.de/landesdenkmalamt/veranstaltungen/denkmalpreis/bundespreis-handwerk/artikel.652230.php.

De La Warr Pavilion

Bexhill-on-Sea, United Kingdom
Erich Mendelsohn and Serge Chermayeff, 1935

Rear (south) facade of the De La Warr Pavilion during construction, Bexhill-on-Sea, United Kingdom, 1935.

Structural steel and steel frame window wall at the south staircase during construction, 1935.

The De La Warr Pavilion, located in Bexhill-on-Sea, is one of the first public buildings in the UK designed in the "International Style," whose principles were laid out in 1932 at the *Modern Architecture: International Exhibition* held at the Museum of Modern Art in New York City. Designed by Erich Mendelsohn and Serge Chermayeff, it was the first building in Britain to employ a welded-steel structural frame as opposed to reinforced-concrete.[1] Its existence is owed to the 9th Earl De La Warr, who wanted to establish a center for arts and entertainment at the seaside town of Bexhill-on-Sea (of which he was mayor at the time). His initiative prompted an international competition organized by the Royal Institute of British Architects (RIBA) in 1933. The design brief did not specify any particular style but petitioned for a design that was simple and used large windows, terraces and canopies (suggestive of a lightweight building). Mendelsohn and Chermayeff's winning entry included a hotel, a swimming pool, a pier and a two-story pavilion with an auditorium. While only the latter two features were actually built, their use of steel and concrete in the design introduced a Modern architectural vocabulary aimed at increasing access to cultural events and leisure for the local residents.[2] Accord-

Exterior detail of the existing steel frame single-glazed window wall at the south staircase, 2018.

Interior detail of the existing steel frame single-glazed window wall at the south staircase, 2018.

ing to *The Times*, it was "by far the most civilized thing that has been done on the South Coast since the days of the Regency."[3]

The pavilion has an east-west linear composition with two wings connected by a prominent circulation core accentuated by glazed staircases at each end. The south facade, overlooking the English Channel, has one of the most striking features of the building design—a helical three-story stair with polished chrome handrails enclosed in a curved steel frame single-glazed window wall with bent glass. The south stair window wall is constructed with multi-story non-thermally broken flat steel mullions and curved flat transom plates (also non-thermally broken and with the same detailing and sightlines as the mullions). The large single-pane bent glass lites have exterior glazing. The interior side of the existing mullion and transom plates are sandwiched with supplemental steel plates in an apparent attempt to stiffen them, although field evidence suggests that this reinforcement is not an original condition. The stair on the north facade is more modest in size, cantilevered above the main entrance to the building, and enclosed by a two-story curved steel frame single-glazed window wall. The north stair enclosure is made with non-thermally broken standard T-shaped steel mullions and fixed steel frame windows with T-shaped muntins and 1/4" (6mm) glass set in putty. At the interior of the south stair window wall, the vertical linear steam tube radiators attached to the mullions are original to the building. They are connected at the top and together create three heating loops that are elegantly integrated into the design of the window wall. A similar set of horizontal tube radiators is located at the north stair enclosure.

TIMELINE

1933	Design competition organized by RIBA
1934	Mendelsohn and Chermayeff winning entry declared
1935	Construction of existing building completed
1960–1970s	Window replacement with wooden frames
1971	Designated a Listed Building by the Minister of Housing and Local Government
1986	Listed as Grade I structure by English Heritage (now Historic England)
1989	De La Warr Pavilion Charitable Trust set up to safeguard the building
1992	John McAslan & Partners appointed to create a strategy for the pavilion's long-term future use
1993–2006	Phased rehabilitation (facade repairs, entrance renovation, new ground-floor gallery and upper level café

Interior view of the existing steel frame single-glazed window wall at the north staircase, 2018.

Interior view of the south staircase window wall, 2018.

Existing west and south facades, 2018.

At the front (north) facade, the east wing has steel frame ribbon windows at both levels, whereas the west wing has three pairs of out-swing double doors at the ground floor level of each bay, surmounted by a cantilevered eave that supports the lettering announcing the pavilion's name. The original steel frame glazed assembly for the north facade of the east wing (where the café, library and first floor bar were located) consisted of standard W20 steel sections with sliding doors manufactured by Crittall Manufacturing Company Ltd.[4] Phosphor bronze draft excluders were used for weatherproofing the assembly. The rest of the window openings in the building were glazed using steel sections. The 1/4" (6mm) thick plate glass was supported by mullions and transoms on the south facade.[5]

At the south facade, the two-story east wing is divided into ten bays. Nearly all of the bays are enclosed by a window wall with sliding doors that used to open the interior towards a ground floor terrace and a second floor balcony overlooking the ocean. On the upper levels, the cantilevered balcony and eaves follow the curved glazing plane of the south staircase and provide sun shading. The three-story west wing (which houses the auditorium) has six glazed bays on the first level, two of which have out-swing casement windows and the other four bays have three pairs of out-swing doors.

During the building design phase, Mendelsohn and Chermayeff sought advice from the Building Research Station, a division of the Ministry of Public Building and Works, "on any new form of paint or anti-corrosive process before painting steel windows, such as, for example, zinc-spraying."[6] They were also interested in using both bronze and aluminum for the windows and doors as a corrosion-resistant alternative to painted steel. For

Interior view of the south staircase, window wall and steam loop pipes, 2018.

Interior view of the south staircase window wall and large pendant light fixture, 2018.

cost-saving reasons none of these were implemented.[7] However, the fact that they were considered indicates that Mendelsohn and Chermayeff were interested in mitigating any corrosion resulting from exposure to the site's maritime environment. The pavilion construction was completed within a year and opened to the public in December 1935. The sense of openness and use of exquisite finishes along with an elegant concept of flowing spaces was widely praised by the visitors.[8]

CONDITION PRIOR TO INTERVENTION

The pavilion was constructed primarily by unskilled laborers, which is thought to be one of the contributing factors leading to the extent of the decay the building suffered over the years.[9] During WWII the building was used for civilian and military administration, and in 1940 it suffered bomb-related damages (primarily to the auditorium).[10] Post-war, it was subjected to various low-quality alterations and additions that concealed its original massing and configuration.[11] The existing reinforcing sandwiched steel plates at the south staircase window wall are likely to have been installed during this period.

Decades of constant exposure to a high-humidity environment, water infiltration and a lack of expansion joints led to the crumbling of the exterior stucco and corrosion of the welded steel structure. Many original elements were lost due to decay, includ-

Interior detail of the south staircase, window wall and steam loop pipe, 2018.

Corroded steel frame mullion at the south staircase, 2018.

ing most of the original steel frame windows.[12] Subsequent replacement windows, like many of the 1930s fixtures and fittings that were installed, were inadequate and inappropriate. Later on, the corroded steel frames of the sliding glazing on the south elevation were replaced with wooden frames, which affected the relationship between the internal and external spaces. Although the glazing of the stair enclosure was consistent with the original design, broken glass panes and deteriorated framing sections were replaced routinely as a maintenance measure up to 2005.[13]

Despite the ongoing decay, the De La Warr Pavilion was given the status of a listed building by the Minister of Housing and Local Government in 1971. This designation was followed by a listing as a Grade I structure in 1986.[14] In 1989, local residents formed a committee for the future maintenance of the Pavilion. This local committee, later known as the De La Warr Pavilion Charitable Trust, has been instrumental in its preservation.

Ongoing corrosion at the bottom of the existing steel frame single-glazed window wall mullion at the south staircase, 2018.

INTERVENTION

The restoration campaign consisted of a multi-phase design and build project under the leadership of John McAslan and Partners.[15] The scope of work included removing the various alterations that over the years had compromised the integrity of the original design intent and had obliterated the prominence of the building's Modernist expression. Non-original finishes were removed throughout the interior and replaced with white-painted wall finishes that allude to the original plaster finishes. The building program was changed to include the current gallery on the ground floor of the east wing (at the location of the original café and bar) and a café on the second floor (at the location of the former original library and reading room). The auditorium was renovated and accessibility was improved by cutting a new ramp through the

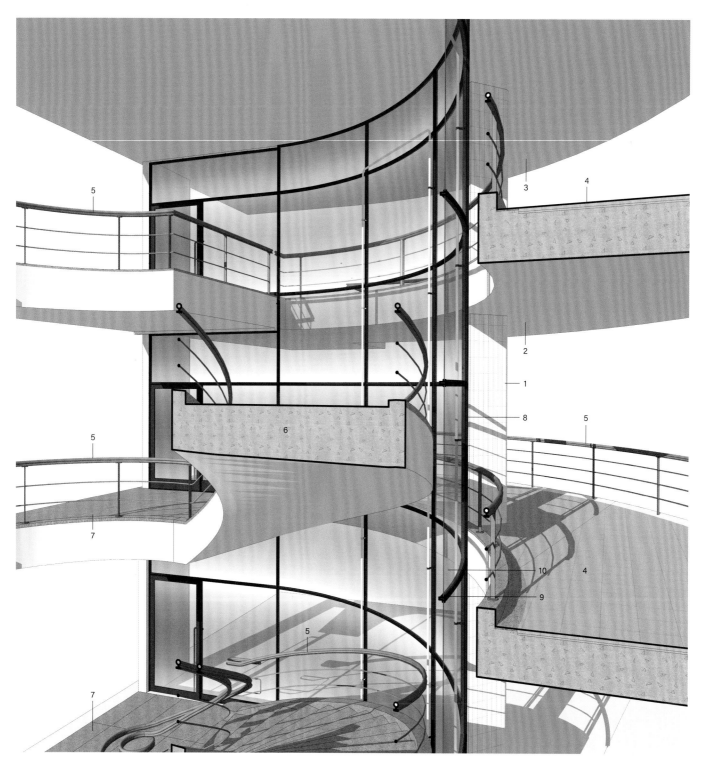

Typical section at south staircase steel frame window wall with bent glass.

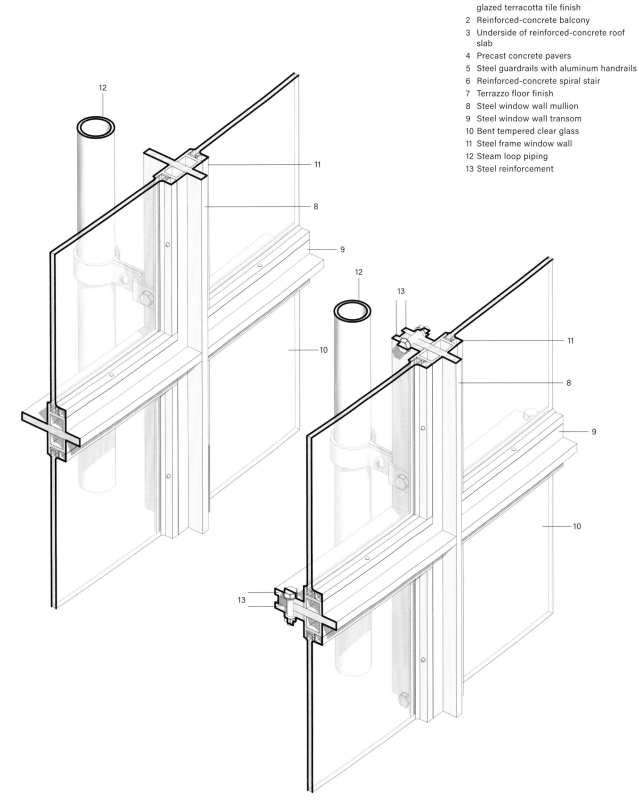

1 Concrete-encased steel columns with glazed terracotta tile finish
2 Reinforced-concrete balcony
3 Underside of reinforced-concrete roof slab
4 Precast concrete pavers
5 Steel guardrails with aluminum handrails
6 Reinforced-concrete spiral stair
7 Terrazzo floor finish
8 Steel window wall mullion
9 Steel window wall transom
10 Bent tempered clear glass
11 Steel frame window wall
12 Steam loop piping
13 Steel reinforcement

Typical before (left) and after (right) window wall details.

middle bay between the two sets of handrails on the stair leading from the entrance level to the ground floor.

Sliding non-thermally broken aluminum frame doors with IGUs were installed at one bay of each floor of the south facade, thereby reestablishing access to the south terrace and the second-floor balcony (albeit with a different material). During the restoration, a non-thermally broken galvanized W20 steel frame window and door system with single-pane tempered glass was installed at the stair and ribbon window locations requiring replacement. The same system was used for the glazed enclosures and doors at the north and south facades of the west wing and at the north main entrance. The glazing was sealed using fin-seal weatherstripping and neoprene buffers.[16] The glazing and exterior finish specifications were aimed at ensuring the restoration work resembled the original intentions of the building's architects. As part of the renovation, the steel plates reinforcing the steel mullions and transoms on the south staircase were resecured with exposed hex bolts. The linear steam tube radiators on the interior side of the south stair mullions and along the north stair transoms were cut out and left in place although disconnected at the bottom from the supply steam source.[17] At the north stair enclosure, field evidence indicates that all the glass panes were replaced with flat heat-soaked glass panes by Pilkington. The steel frame windows were painted with Tema Antiruggine (a water-soluble anti-corrosion paint, manufactured by Ard Raccanello).[18]

COMMENTS

The De La Warr Pavilion is an unexpected burst of Modernity amid the Edwardian and Victorian architecture that predominantly defines the character of the seaside town of Bexhill-on-Sea. The window walls, particularly at the glazed stair enclosures, define its Modernism as much its morphology does—with long wings and ribbon windows contrasting against white walls, cantilevered balconies and eaves. At the north facade, where the window wall frames a picturesque view of the town from the staircase, the wall becomes an ethereal diaphragm between the town's past and the modern present dreamed of by the 9th Earl De La Warr. This is why the replacement flat glass is extremely disappointing, for when looked at closely it is obvious that it is fit poorly into the bent steel muntins. Also somewhat disappointing are the non-functioning steam tube radiators at the transoms of the north and south stair mullions that were disconnected and rendered obsolete without implementing a suitable alternative solution to prevent heat loss and interior condensation during the winter. Fortunately, this is one condition that could be reversed during a future intervention.

From the start, the corroding effect of its seafront environment has been one of the most challenging design and preservation issues for the building. Mendelsohn and Chermayeff's inquiries

Ongoing corrosion at the existing replacement steel frame door and windows at the south staircase, 2018.

Existing steel frame window wall with broken bent glass at the south staircase, 2018.

West view of the south staircase, 2018.

East view of the south facade, 2018.

CREDITS

De La Warr Pavilion (1935)
Bexhill-on-Sea, United
Kingdom

ORIGINAL CONTRUCTION

Architects

Erich Mendelsohn and
Serge Chermayeff

Structural Engineer

Felix Samuely

Contractor
Rice & Sons (from Brighton)

Structural Steel
Helsby, Hamann and Samuely

1993–2006 INTERVENTIONS

Owner
Rother District Arts Council

Property Manager
De La Warr Pavilion Charitable
Trust (2002–Present)

Architects
John McAslan + Partners

Structural Engineer
F. J. Samuely and Partners Ltd.

M&E Engineers
Rybka

Cost Consultants
Maynard Mortimer & Gibbons

Contractor
Heasman Spicer

South facade, 2018.

about other corrosion-resistant materials for the glazed enclosures, such as bronze and aluminum, indicates their understanding of the challenges posed by the building environment. Availability and cost aside, the use of these softer materials, far less rigid than steel, would have at minimum required heavier framing that was incompatible with the intended lightness of their design. That is why these substitute materials are not a viable option for future interventions.

The ongoing breakage of the large bent glass and at other locations, which took place after the restoration work completed in 2006, is clearly related to corrosion of the steel framing on the window walls and poses aesthetic and safety challenges. These assemblies require a reconsideration of materials, finishes and preventive actions to mitigate the effect of exposure to mist and sea salt, which are the leading causes of the underlying corrosion. To meet this challenge, a conservation program must be established that combines archival research with thorough assessment of field evidence and a well-defined conservation approach, to clearly outline where repairs or alternative measures are suitable and appropriate, and where replacement is required. The De La Warr Pavilion Charitable Trust is currently pursuing a conservation plan that will hopefully accomplish these goals.

NOTES

1   Richard Carr, "De La Warr Pavilion," Studio International, last modified 14 November 2005, accessed 13 January 2019, http://www.studiointernational.com/index.php/de-la-warr-pavilion.

2   Mark Cannata, "The Repair and Alterations of the De La Warr Pavilion," Journal of Architectural Conservation 12, no. 2 (2006): p. 81, DOI: 10.1080/13556207.2006.10784970.

3   Modernist Britain, "De La Warr Pavilion, Bexhill-on-Sea," last modified 1 January 2014, accessed 13 January 2019, http http://www.modernistbritain.co.uk/post/building/the+de+la+warr+pavilion/.

4   Allen Cunningham, Modern Movement Heritage (London: E & FN Spon, 1998), pp. 133–134.

5   Ibid.

6   Alastair Fairley, De La Warr Pavilion: The Modernist Masterpiece (London/New York: Merrell Publishers, 2006), p. 78.

7   Fairley, De La Warr Pavilion: The Modernist Masterpiece, p. 122.

8   Cannata, "The Repair and Alterations of the De La Warr Pavilion," p. 81.

9   Ibid.

10 Modernist Britain, "De La Warr Pavilion, Bexhill-On-Sea."

11 Cannata, "The Repair and Alterations," p. 88.

12 Ibid.

13 Cunningham, Modern Movement Heritage, p. 134.

14 Modernist Britain, "De La Warr Pavilion, Bexhill-On-Sea."

15 Sean Albuquerque (board member, De La Warr Pavilion Charitable Trust; ABQ Studio Architects), in on-site discussion with Angel Ayón, 22 August 2018.

16 Cunningham, Modern Movement Heritage, p. 134.

17 Cannata, "The Repair and Alterations," p. 85, fig. 4.

18 Rupert Harris Conservation, letter to Sean Albuquerque, 22 May 2012.

Lever House

New York, New York, USA
Gordon Bunshaft, Skidmore, Owings and Merrill, 1952

Lever House as viewed from Park Avenue in New York, New York, ca. 1952. © Ezra Stoller/Esto

Interior view of south-facing curtain wall, 1952. © Ezra Stoller/Esto

Designed by Gordon Bunshaft from the New York office of Skidmore, Owings and Merrill (SOM), Lever House, at 390 Park Avenue in Manhattan, was the first all-glass International Style office building in the US. Although the United Nations Secretariat building (Wallace K. Harrison, Le Corbusier, Oscar Niemeyer, with an international consortium of architects, 1950) was the first to exhibit a Modern curtain wall (made of aluminum frames), it was Lever House that was the first to expand the concept to all four facades of the tower (except at the lower floors of the west facade, where there is a masonry party wall). The curtain wall at Lever House, which has no operable windows, includes the tower, as well as the second floor of the building's base both at the street facades and at the courtyard. Before it was completed, *The New Yorker* praised the ingenious solution devised by SOM to clean the all-glass facades by means of a revolutionary custom-made "electrically operated platform." Designed by Otis Elevator Company, the system predated the contemporary window washing scaffold now common

Closer view of the south facade after the curtain wall replacement, 2019.

View of base from the courtyard, 2007.

View from the Seagram Building on Park Avenue after the curtain wall replacement, 2019.

to most high-rise buildings (it also included a phone line so that the operator could communicate with the building engineer's office).[1] The emphasis on maintaining a sparkling clean building with minimum difficulty or expense made total sense for Lever Brothers, a company that specialized in manufacturing cleaning products such as soaps and detergents. In a follow-up article in *The New Yorker* published shortly after the building opened in 1952, one of the Lever brothers remarked that they wanted their building to "be a symbol of everlasting cleanliness [given that they were in the soap business]" and that meant, "a building where not just the windows but every inch of surface could be washed regularly."[2]

Set back 100' (13.5m) from the south property line and 40' (12.19m) from the north, the four-sided glass-clad steel frame tower of Lever House is 60' (18.29m) wide and 20 stories high, rising above a two-story base with a central court. The curtain wall is made of concealed steel members with 16-gauge (1.3mm) Type 302 stainless steel covers secured to the exterior glazing channels

TIMELINE

1950–1952	Completed
1982	Designated a New York City Landmark
1983	Listed in the National Register of Historic Places
1998	Purchased by RFR Holding LLC, the current owner
2001	Restoration completed

with hand-driven screws.[3] The original heat resistant glazing (spectrally selective tinted glass) consisted of a single lite of greenish-blue vision glass at the 4'–6" (1.37m) wide fixed windows and two lites of bluish-green wire glass at the spandrels. When Lever House opened in 1952 to grand fanfare, it was dubbed "the house of glass" by *The New Yorker*.[4] In the ensuing years, it redefined the commercial office building typology, unleashing a new urban paradigm of buildings set back from the street, rising from atop a horizontal base and lifted off the ground plane by pilotis. Its all-glass curtain wall with stainless steel mullion and transom covers became the epitome of Modern elegance, urban sophistication and corporate pretentiousness.

CONDITION PRIOR TO INTERVENTION

The building's blue-green glass facade had deteriorated due to harsh weather conditions and the limitations of the original fabrication and materials. Water had seeped behind the stainless steel mullion covers causing the carbon steel within and around the glazing pockets to rust and expand. This corrosion bowed the transom bars and broke most of the glass on the spandrel panels. The corrosion was barely visible from the thoroughfare however, as it remained hidden behind the stainless steel covers.[5] By the late 1990s, very little of the original glass remained. According to Aby Rosen from RFR Holding (the building owner since 1998), the curtain wall had deteriorated significantly and "it was like some of the glass was just hanging on little strings of metal that was corroded away."[6]

INTERVENTION

Due to the poor condition of the curtain wall, the initial plan for the restoration, which included localized glass replacement and steel frame repairs, had to be abandoned. Rosen and SOM (who spearheaded the restoration) reportedly agreed not to attempt to salvage the existing stainless steel and glass.[7] When discussing the choice to replace all the stainless steel covers rather than repair them, SOM principal David Childs explained that the firm's approach was based on the assumption that "what was important about this [building] is that it was machine-made." He further described the original building as follows:

> It has smoothness and a machine quality to it. And it was better in the preservation of it to throw away the dented pieces of the structure. To actually take a new piece of stainless steel—exactly the same technology and fabrication—and replace that piece. That would be a better way to do it than to save this battered piece— which could never be made to look machine-made again.[8]

Base and tower as viewed from Park Avenue, 2019.

View from the sidewalk on Park Avenue, 2019.

As part of the intervention, all the glass was removed and re-placed with heat-strengthened spectrally selective tinted glass Solex by PPG.[9] The deteriorated steel sub-frame was repaired where possible or replaced with concealed aluminum glazing channels at deteriorated locations. The two-panel glass spandrels, whose original dimensions were limited by the size of the wired glass available at the time of the original construction, were re-placed with larger single-glazed laminated glass units with a non-functioning matching transom to retain the original appear-ance. This reduced the materials and installation costs as well as the total number of joints, thus affording enhanced moisture pro-tection. Its status as a New York City Landmark and its listing in the National Register of Historic Places allowed the building to be exempt from compliance with energy conservation requirements, which meant the original single-glazed configuration that defines the building's architectural significance could be preserved. All of the original stainless steel mullion covers were replaced with matching versions, as it was deemed too costly to remove, catalog and store them during the course of the work.

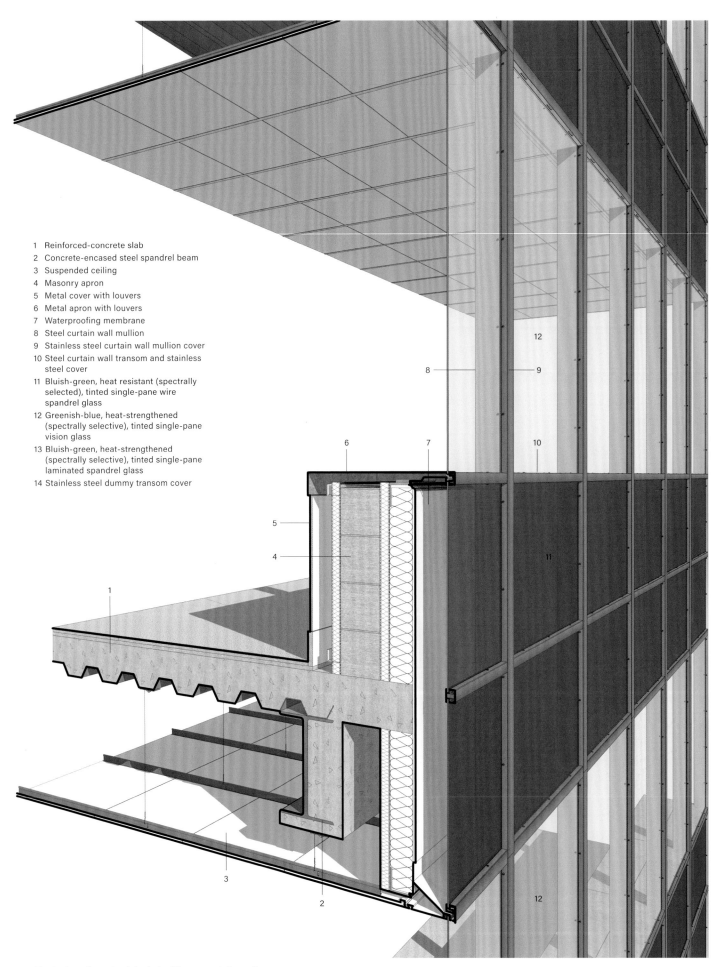

1 Reinforced-concrete slab
2 Concrete-encased steel spandrel beam
3 Suspended ceiling
4 Masonry apron
5 Metal cover with louvers
6 Metal apron with louvers
7 Waterproofing membrane
8 Steel curtain wall mullion
9 Stainless steel curtain wall mullion cover
10 Steel curtain wall transom and stainless steel cover
11 Bluish-green, heat resistant (spectrally selected), tinted single-pane wire spandrel glass
12 Greenish-blue, heat-strengthened (spectrally selective), tinted single-pane vision glass
13 Bluish-green, heat-strengthened (spectrally selective), tinted single-pane laminated spandrel glass
14 Stainless steel dummy transom cover

Typical section at original steel frame curtain wall.

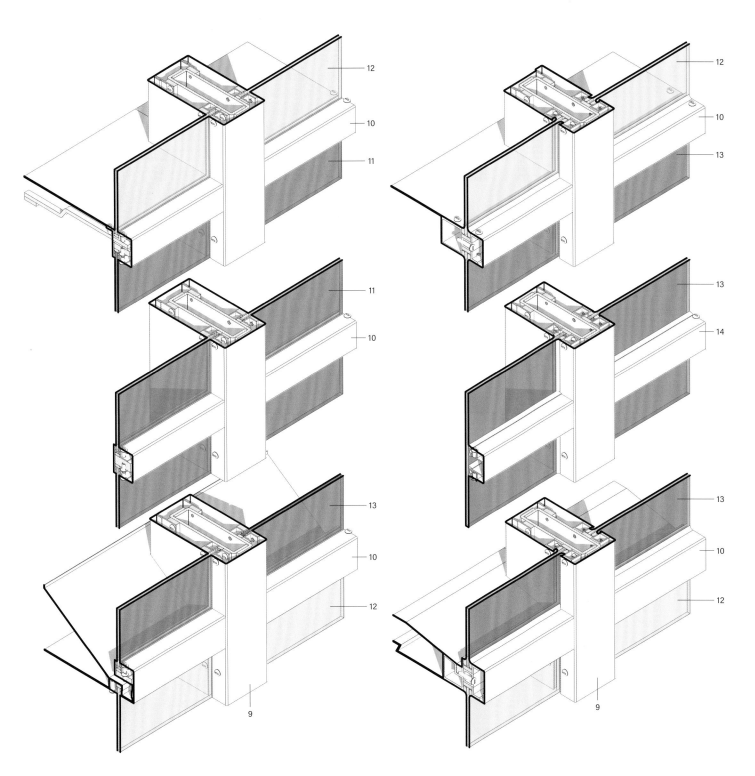

Typical before (left) and after (right) curtain wall details.

Looking up from the sidewalk on Park Avenue, 2019.

COMMENTS

At the renovated Lever House, the new stainless steel mullion covers, coupled with the replacement of all vision and spandrel glass, lent the building an as-new appearance that makes it one of the most refreshing facades on Park Avenue. The most drastic changes on the facade are concealed, and therefore imperceptible. The new tinted glass and replacement stainless steel covers returned to Lever House the appearance of "everlasting cleanliness" that the original owners sought when they commissioned the building more than half a century ago.

The replacement of the original glass with new single-pane glass, however, is a choice that deserves scrutiny. While discussing the renovation of Modern curtain wall buildings, Gordon

CREDITS

Lever House (1952)
New York, New York

ORIGINAL CONSTRUCTION

Owner
Lever Brothers

Architects
Gordon Bunshaft, Skidmore,
Owings & Merrill (SOM)

Interior Design
Raymond Loewy Associates

Structural Engineers
Weiskopf & Pickworth

Mechanical Engineers
Jaros, Baum & Bolles

Contractor
George A. Fuller Company

1998–2001 INTERVENTIONS

Owner
RFR Holding

Architects
Skidmore, Owings & Merrill
LLP (SOM)

Consulting Engineer
Vincent Stramandinoli

Exterior Wall Consultant
Gordon Smith, PE, Gordon H.
Smith Corporation

Interior Architect
William T. Georgis Architect

Graphic Design
Pentagram

Landscape Design
Ken Smith, Ken Smith
Workshop

Curtain Wall Contractor
Flour City Architectural Metals

Base and tower as viewed from East 53rd Street, 2019.

Smith, the exterior wall consultant responsible for the intervention on the Lever House curtain wall, called for "faithful replication" of the facades of significant Modern buildings. He added that he favors replacement that reinstates the original appearance, though not of all materials or performance.[10] Intuitively, this seems reasonable for Lever House, especially regarding the vision glass, where retention of the original historic character is desirable for this highly significant curtain wall assembly. However, at the spandrels, which account for roughly half the area of the curtain wall, retaining the original single-pane glass configuration instead of replacing it with IGUs was a missed opportunity for improving the building's thermal performance without impacting its historic appearance and significance. At Lever House, the perimeter concrete-encased steel beams, reinforced-concrete slab edge and metal apron in front and above the radiators conceal the inboard side of the spandrels. That means that the spandrels are only visible from the exterior and never from the interior. They could only be exposed from the interior by disassembling the aprons. Therefore, a faithful replication and reinstatement of the original appearance would have been attainable, all while reducing the likelihood of interior condensation—a key deteriorating factor for steel frame curtain wall assemblies—and minimizing energy loss through the single-pane glass. This approach though seems to have been disregarded because the historic building did not need to comply with energy conservation requirements. This is a misinterpretation of some of the main tenets of historic preservation, where appropriateness is ideally based on how each intervention balances materials conservation, systems performance and retention of the historic image, rather than just satisfying one at the expense of the others.

NOTES

1 "Solution," *The New Yorker,* 26 May 1951, The Talk of the Town section, p. 21.

2 "Clean," *The New Yorker,* 26 April 1952, The Talk of the Town section, p. 27.

3 "Lever House, New York: Glass and Steel Walls," *Architectural Record* 111 (June 1952): p. 131.

4 Lewis Mumford, "House of Glass," *The New Yorker*, 9 August 1952, The Sky Line section, p. 48.

5 "Landmark's Curtain Wall Is Deteriorating; Aging Lever House May Get a New Skin," *The New York Times,* 22 October 1995, accessed 23 December 2018, 2018, https://www.nytimes.com/1995/10/22/realestate/postings-landmark-s-curtain-wall-deteriorating-aging-lever-house-may-get-new.html.

6 Matt Tyrnauer, "Forever Modern," *Vanity Fair,* October 2002, accessed 23 December 2018, 2018, https://www.vanityfair.com/culture/2002/10/leverhouse200210.

7 Ibid.

8 Ibid.

9 SOM, "Projects – Lever House – Curtain Wall Replacement," accessed 23 December 2018, https://www.som.com/projects/lever_house__curtain_wall_replacement.

10 Angel Ayón and Nina Rappaport, "Greening the Glass Box: A Roundtable Discussion about Sustainability and Preservation of Modern Buildings," *MÓD* issue 1 (DOCOMOMO New York/Tri-State, 2014): p. 19.

S. R. Crown Hall

Illinois Institute of Technology (IIT), Chicago, Illinois, USA
Ludwig Mies van der Rohe, 1956

Front (south) facade of S. R. Crown Hall at Illinois Institute of Technology (IIT) in Chicago, Illinois, during construction, ca. 1956.

East facade during construction, ca. 1956.

Crown Hall is one of the greatest architectural achievements of German-American architect Ludwig Mies van der Rohe (1886–1969). Its clear-span steel-and-glass construction epitomizes the Modernist quest for structural clarity, material innovation and adaptability of interior space. The ingenious use of four inverted 6' (1.8m) deep rooftop steel girders supported by perimeter columns enabled the first large-scale realization of Mies' concept for a flexible "universal space" within a building.[1]

The exterior appearance of this two-story educational facility is dominated by the presence of the black exposed beams and perimeter steel columns that brace the steel mullions of both the first floor window wall and the basement windows. At the first floor, which holds the main hall with the design studios, the window wall bays between columns are subdivided into three fixed lites and an operable grille at the bottom. These fixed lites are separated by a mullion and a transom with rectangular glazing beads made of stock bars detailed similarly to those at the Farnsworth House in Plano, Illinois (see page 82).

The original first-floor glazing at each window bay above the operable grilles included two fixed lites with 1/4" (6mm) thick sandblasted glass that are 7'–9" (2.36m) high, and one 9'–7" (2.92m) wide by 11'–6" (3.51m) tall lite above with 1/4" (6mm) thick clear polished plate glass. The clerestory hopper windows at the

Colorized photo shortly after construction, 1956.

Detail of main (south) entrance, 1956.

basement level, located between the top of the foundation wall and the exposed first-floor steel beam, are 4' (1.22m) tall with the same detailing, width and sandblasted glass as the lower panes of the first-floor window wall above. This modular configuration and combined use of sandblasted and clear glass is typical throughout the facades except at the entrances. Located on the north and south facades between the two middle girders, the entrances only include clear glazing. It is here and through the main hall and the design studios, and especially around the window wall, where Mies' use of the building structure and the steel frame glazed enclosures to create the illusion of a weightless and adaptable universal space is most palpable.

CONDITION PRIOR TO INTERVENTION

Mies was dismissed as IIT campus architect in 1958, after which the academic building that he designed continued to be used intensively for its original purpose.[2] The original 1/4" (6mm) thick clear glass was too thin for the large upper lites at the main level and broke often. The interior of the sandblasted glass at the lower lites in the main level had become heavily soiled, as the rough profile of the sandblasted finish attracted all manner of dirt, oils and adhesive residues from student fingerprints, model-making and general use as a pin-up wall. After almost two decades of continued wear and tear, all of the original 1/4" (6mm) glass was removed and replaced in 1975, following the recommendations of Skidmore, Owings and Merrill (SOM), who succeeded Mies as campus architect. After evaluating the building, SOM concluded that the original glass did not meet Chicago's updated building code and was inadequate to resist lateral (wind) loads. The firm also determined that the original steel glazing stops were poorly

TIMELINE

1950–1954	Design period
1954–1956	Building construction
1975	Exterior glazing and roof are replaced, interior modifications completed
1985–1986	Repairs to exterior porches and doors are made, air conditioning installed
1989–1996	Basement renovation and roof replacement completed, air conditioning modifications made
1997	Designated as a Chicago Landmark
2001	Listed in the National Registry of Historic Places as a National Historic Landmark
2002	Krueck and Sexton Architects selected to undertake project for major renovation of Crown Hall
2003	Master plan by Krueck and Sexton Architects and others is completed
2005–2006	Restoration of window wall is completed

Interior view of studio space, 1956.

Detail view showing top of replacement transom bar, 2016.

detailed and that the glazing compounds had hardened and no longer cushioned the glass from the steel. In response to these findings, the upper clear glazing was replaced with 3/8" (9.5mm) thick clear glass, installed with replacement painted aluminum glazing stops. While minimal, the new stops did create a thicker sightline than the original steel mullions. At the lower lites, the sandblasted glass was replaced with 7/32" (5.6mm) translucent glass, installed into the original steel frame and glazing stops using a contemporary wet-seal silicone glazing compound.[3]

In 1996, replacement of the exterior glazing was implemented once again. By then, the steel framing at the main (south) entrance terrace had severely deteriorated, and Chicago's building code had been updated. The extant glazing thicknesses were then deemed inadequate for wind loads, and safety glazing was also required for the lower lites at the first floor.[4] By 2001, all the clear upper lites at the main level had been replaced with 1/2" (12.7mm) clear annealed glass, and the lower clear lites at the entrances had been replaced with 1/4" (6mm) tempered glass, which provided safety protection from impact breakage. The replacement translucent glass at the lower lites had been replaced with 1/4" (6mm) laminated glass with a mylar interlayer to simulate the translucent qualities of the original sandblasted glass.[5] This intervention eliminated the ongoing staining, but the result was more reflective than Mies' original matte finish.[6]

INTERVENTION

In 2002, IIT retained a team spearheaded by Krueck and Sexton Architects, McClier (which became Austin/AECOM) and other consultants to safeguard the building's seminal status. The resulting facade work entailed window wall restoration and rehabilitation, including removal of existing glazing and steel stops, removal

Close-up view of replacement transom bar, laminated clear glass and restored mullion at rear (north) entrance, 2016.

Replacement glass and steel glass stops at the restored steel frame window wall, 2016.

Interior view of studio space showing the window wall after the restoration, 2015.

of all lead-based paint from interior and exterior steel surfaces, as well as cleaning, repair and painting of corroding steel. The original ventilation grilles at the bottom of the window wall were also cleaned, refurbished and provided with new electromagnetic hardware. Contemporary requirements for overhead glazing would not allow the in-kind replacement of the 1/4" (6mm) thick clear upper glass lites. Therefore, the challenge of reglazing Crown Hall was about reconciling the design intent and bygone appearance of the original glass with contemporary requirements for glass safety.

One code-compliant option for the large panes required for the upper lites might have been to replace the original 1/4" (6mm) thick polished plate glass with 3/8" (9.5mm) tempered glass. This option, however, was discounted due to the slight light distortions that characterize tempered glass, which are the result of rolling the sheet glass over steel rollers during the tempering process.[7] Instead, the replacement choice for the upper lites at the main level was 1/2" (12.7mm) thick Starphire low-iron glass by PPG, installed in custom-made glazing stops with a tapered end to provide the required glass setting while retaining the original appearance. The heavier 1/2" (12.7mm) thick glass required the installation of setting blocks to accommodate deflection of the

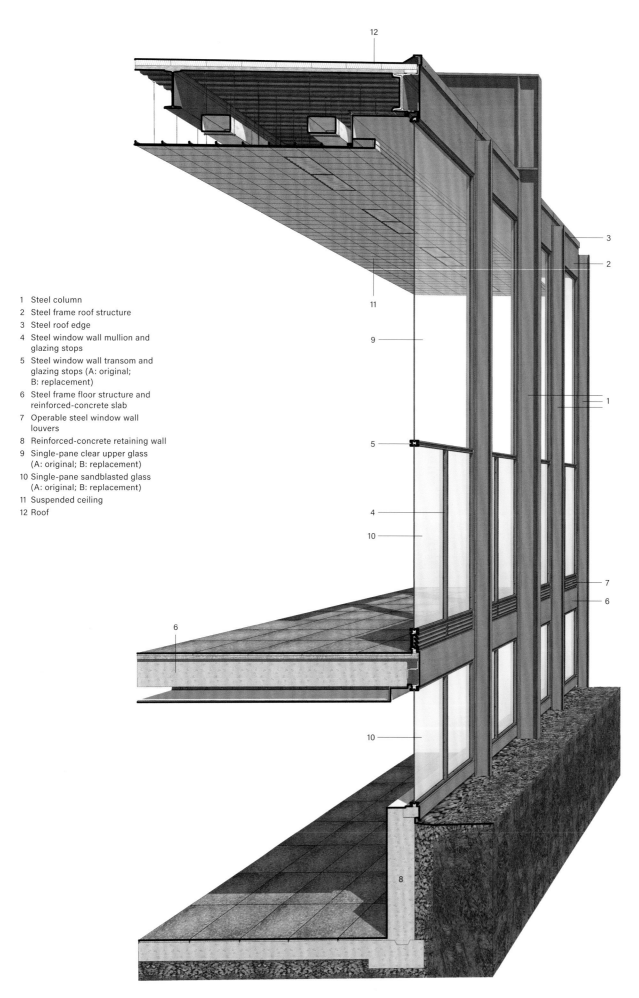

1 Steel column
2 Steel frame roof structure
3 Steel roof edge
4 Steel window wall mullion and
   glazing stops
5 Steel window wall transom and
   glazing stops (A: original;
   B: replacement)
6 Steel frame floor structure and
   reinforced-concrete slab
7 Operable steel window wall
   louvers
8 Reinforced-concrete retaining wall
9 Single-pane clear upper glass
   (A: original; B: replacement)
10 Single-pane sandblasted glass
   (A: original; B: replacement)
11 Suspended ceiling
12 Roof

Typical section at original steel frame window wall.

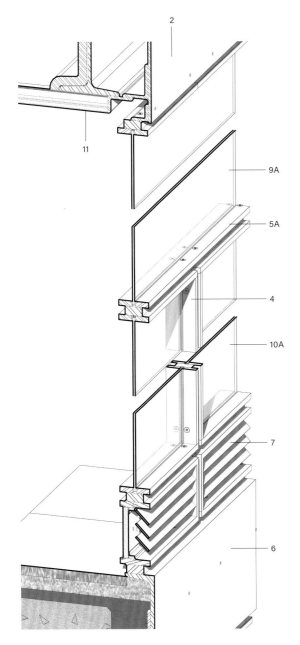

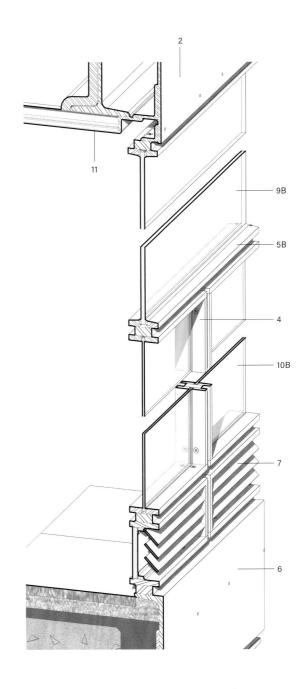

Typical before (left) and after (right) window wall details.

East facade after the restoration, 2016.

glass, a minimum edge clearance of 1/8" (3mm) and a minimum glass bite of 1/2" (12.7mm).[8] As these dimensions could not be accommodated by the original 5/8" (16mm) high glazing stop or by the original 1/2" (12.7mm) glazing bite, a new 3/4" (9mm) steel glazing stop profile was custom-made with a tapered top pitched towards the exterior and of a height matching the original 5/8" (16mm) dimension. The lower portion of the window wall was replaced with tempered 1/4" (6mm) sandblasted safety glass by Viracon, coated with ultra-clear epoxy in order to maintain the sandblasted appearance while preventing staining and scratching.[9] The steel work was painted with an epoxy coating by Tnemec matching the original black color.

COMMENTS

The refurbishment and restoration of Crown Hall is an audacious and sensitive intervention that restored the visual integrity of the original design and made it perform more efficiently and safely.

View showing faded coating at restored steelwork on the southwest corner, 2016.

CREDITS

S. R. Crown Hall (1956)
Chicago, Illinois

ORIGINAL CONSTRUCTION

Architects
Ludwig Mies van der Rohe, David Haid (project architect from Mies' office), Myron Goldsmith (Mies' former student)

Consultants
Alfred Caldwell (Landscape/ Garden Designer)

Frank Kornacker and Associates (Consulting Engineers)
PACE Associates (Construction documents and management)
Dahl-Stedman Company (Contractor)

1975–1996 INTERVENTIONS

Architects
Skidmore, Owings and Merrill (1975)
George Schipporeit, Peter Beltemacchi, David Sharpe (1985–1986)

Gene Summers, Fujikawa Johnson and Associates (1989–1996)

2002–2006 INTERVENTIONS

Architects
Krueck and Sexton Architects

Preservation Architects
Gunny Harboe (formerly with McClier Preservation Group, now AECOM)

Environmental Design
Atelier Ten and Transsolar

General Contractor
Clune Construction Co.

Window Wall Subcontractor
Harmon

Interior view of the northeast corner conference room after the restoration, 2016.

The replication of the sandblasted glass finish on tempered glass ensures safety while reinstating the original appearance. However, the protective clear coating does change the visual and tactile properties of the glass, albeit minimally. Inside, now that the rough sandblasted glass finish is no longer exposed, the lower lites are glossy and lack contrast with the clear glazing above and the "Miesian black" on the adjacent steelwork. Without this subtle contrast, the lower half of the curtain wall resembles at times translucent plastic more than a glass surface. This is a case where balancing durability and maintenance against retention of the original appearance has caused the interior experience to be somewhat compromised in favor of a more authentic perception from the exterior. The overall outcome is nonetheless successful.

The change to the replacement glazing stops is imperceptible from either the ground or from inside the main hall. This subtle modification also directs water away from the glazing pocket and matches the original configuration. This was a wise choice that prioritized reinstating the visual characteristics of the original upper lites—the largest of all glass panes—over retaining the materiality and detailing of the original steel frame window wall. At Crown Hall, a minimum change of this nature, imperceptible even to the trained eye (the author could barely notice it), is appropriate so long as it is well-documented and duly incorporated into the presentation of the original building and its subsequent alterations.

The only disappointing issue at Crown Hall is the seemingly poor application of the Tnemec epoxy coating, whose black color has faded considerably toward a charcoal tone on the exterior. Correcting this color deficiency would require repainting more often than needed, or accepting a departure from the original finish shortly after the restoration work was completed. In spite of these shortcomings, Mies' philosophy of "almost nothing" remains quite apparent at Crown Hall.

NOTES

1   Illinois Institute of Technology, and McClier Preservation Group, *S. R. Crown Hall: Historic Structure Report* (Chicago, IL: Illinois Institute of Technology, 2000), p. 7.

2   Mies van der Rohe Society, "Illinois Institute of Technology Master Plan 1939–1958," accessed 12 June 2016, http://www.miessociety.org/legacy/projects/illinois-institute-technology-master-plan/.

3   Illinois Institute of Technology and McClier, *S. R. Crown Hall: Historic Structure Report*, p. 53.

4   Ibid., p. 54.

5   Eric D. Thompson, *National Historic Landmark Nomination: S. R. Crown Hall* (Washington, DC: US Department of the Interior, National Park Service, October 2000), pp. 35-37.

6   Sara Hart, "The Perils of Restoring 'Less is More,'" *Architectural Record* 194, no. 1 (January 2006): p. 151.

7   Ibid., p. 152.

8   Glass Association of North America, *GANA Glazing Manual* (Topeka, KS: Glass Association of North America, 2004), p. 94.

9   DOCOMOMO US Registry, "S. R. Crown Hall, History of Building/Site," accessed 23 December 2018, https://docomomo-us.org/register/s-r-crown-hall.

Convent of La Tourette

Éveux, France
Le Corbusier, 1960

South facade of Convent of La Tourette in Éveux, France, 1960.

In 1952, Father Pierre-Charles-Marie Couturier commissioned Le Corbusier (1887–1965) to design a new convent for 100 Dominican brothers-in-training at a secluded countryside location near Lyon.[1] As one of the brothers later described, his choice was based on "the beauty of the convent to be built, of course, but especially for the significance of this beauty. It was necessary to show that prayer and religious life are not related to conventional forms and that an agreement can be established between them and a more modern architecture."[2] After more than three years of design and a few more of construction, the building was completed in 1959 and inaugurated in the presence of Le Corbusier and Cardinal Gerlier in 1960.

Since its inception, the convent has been celebrated by architectural critics and historians as the last incarnation of Le Corbusier's preference for "massive sculptural shapes in rough concrete that stand at the opposite end of the spectrum from the sleek glass, aluminum, and steel towers so prominent in new city-

Courtyard view towards the *pans de verre ondulatoires* along the ambulatory, 2006.

Courtyard facades showing the *pans de verre ondulatoires* along the ambulatory, 2006.

scapes."[3] The Fondation Le Corbusier describes it as "[a] unique synthesis of the attainments of the Modern Movement, a combination of purist lines, Brutalist surfaces and of exceptional constructive solutions."[4] The use of concrete is such a prominent feature of the building that most of the public spaces have mullions made of it, such as along the *grande conduite* (main walkway) and in the library, refectory, classrooms and ambulatories. Iannis Xenakis' design of staggered glass panes in rhythmic patterns is also incorporated into the fixed floor-to-ceiling glazed enclosures with concrete mullions, thus creating the convent's famous *pans de verre ondulatoires*, or "undulating glazing", on three of the exterior facades and along the inner main walkway.[5] Ventilation is provided by vertical pivoting metal shutters (*aréateurs*) that can be adjusted to different opening positions. This was the first time that this system of fixed glazing and adjustable ventilation openings was applied. Le Corbusier later used it for other buildings, such as the Maison du Brésil (Le Corbusier and Lúcio Costa, 1953–1959) a student housing for Brazilian students and scientists in the Cité Universitaire in Paris, and the Maison de la Culture in Firminy (1961–1965).

CONDITION PRIOR TO INTERVENTION

The glass of the *pans de verre ondulatoires* was kept in place within grooves along the concrete mullions with the aid of glazing putty. Two different *barlotières* (muntins) horizontally separated the original 5/32″ (4mm) thick single glass panes. The higher ones were made of 1″ (25mm) wide H-shaped brass profiles and the smaller ones were made of 1/2″ (13mm) wide inverted U-sections of grey polyurethane.[6] Over the years, the glazing putty along the

TIMELINE

1943	Dominican order buys the La Tourette estate
1953	Le Corbusier is chosen by Dominican Province of Lyon to build a monastery
1959	The brothers move into the convent
1960	Building is inaugurated
1961–1964	Repairs to waterproofing membranes at roof terraces
1964–1979	Repairs of cracks, leakages and waterproofing membranes (not documented)
1970	Complex opens to the public for visits and overnight stays
1979	The convent is listed as a historical monument
1982–1993	First supervised restoration work completed (waterproofing at roofs, treatment of concrete surfaces, replacement of main church doors)
2006–2012	Fire protection measures implemented; west wing restored; south and east wing roofs insulated; *pans de verre ondulatoires* replaced; south and east wing facades and interior spaces restored; skylight fixtures rehabilitated
2013	Restoration is completed

View of the rear (west) facades, 2016.

jambs hardened and cracked. The grey U-shaped profiles hardened, darkened and cracked. The *barlotières* were twisted and deteriorated. Differences in the thermal expansion properties of the glass and the concrete mullions led to occasional glass breakage with the consequential risk of glass panes falling down from a considerable height. In addition to these safety issues, the single-glazed enclosures throughout the building have very poor thermal performance, which increases the cost of heating during the winter and also affects the building thermal mass during the summer.[7]

INTERVENTION

After the convent's inauguration, several alterations and additions ensued, many of which were not thoroughly documented. Up until its listing as a historic property in 1979, most of the repair work focused on repairing the leaking roof terraces. After the building designation, records show that various interventions were undertaken from 1982–1993 under the supervision of Jean-Gabriel Mortamet, Chief Architect of Historic Monuments (Architecte en Chef des Monuments Historiques [ACMH]).[8] In the first stage from 1982–1984, Mortamet addressed the ongoing leaks at the roof terraces by installing a new layer of waterproofing above the existing roof plies, which resulted in a thicker roof construction and the installation of additional lead counter flashing. He also applied a new surface treatment to the exposed concrete surfaces and to the cracks in the parapets above the cells. Later on, these treatments were implemented at other parts of the convent (1985–1987). Between 1989–1993, Mortamet eliminated two existing church doors and replaced them with new ones built according to Le Corbusier's original drawings—an approach that followed Viollet-le-Duc's controversial restoration philosophy of completing an unfinished original design.

The original glazing putty at the *pans de verre ondulatoires* displayed fingernail imprints, 2008.

The *pans de verre ondulatoires* at the Maison du Brésil, where new asymmetric and slightly thicker single-pane glass was installed during the restoration in 2008.

Interior view of the reinforced-concrete mullions and replacement muntins in the refectory, 2014.

A major intervention was undertaken from 2006–2013, which, in addition to the implementation of fire protection measures, included the replacement of the *pans de verre ondulatoires*. The discussions with regard to the intervention on this significant feature centered around technical and aesthetic goals—namely, to achieve watertightness and glass safety while retaining the original glass hue and transparency. Safety concerns made it necessary to replace the original 5/32″ (4mm) thick single-pane glass with 5/16″ (8mm) laminated safety glass (Stadip, manufactured by Saint Gobain).

The inverted U-shaped polyurethane muntins were replaced with translucent plastic H-sections, and the H-shaped brass muntins were replaced with anodized aluminum sections with a glossy, grey finish. This first solution, however, evoked complaints by visitors and experts, who argued that the precision and visual rhythm of the original solution was lost due to the reflections and rounded corners of the new installation. To remedy this, the translucent plastic H-shaped profiles were removed and replaced with custom-made opaque silicone profiles of a light grey tone matching the original, and the non-original aluminum muntins were replaced with new custom-made brass extrusions.[9] The new laminated glass was retained even though it was thicker and did not match the smoothness, color, reflective quality or aging properties of the original.

Interior view of the ambulatory and ramp with the *pans de verre ondulatoires*, 2006.

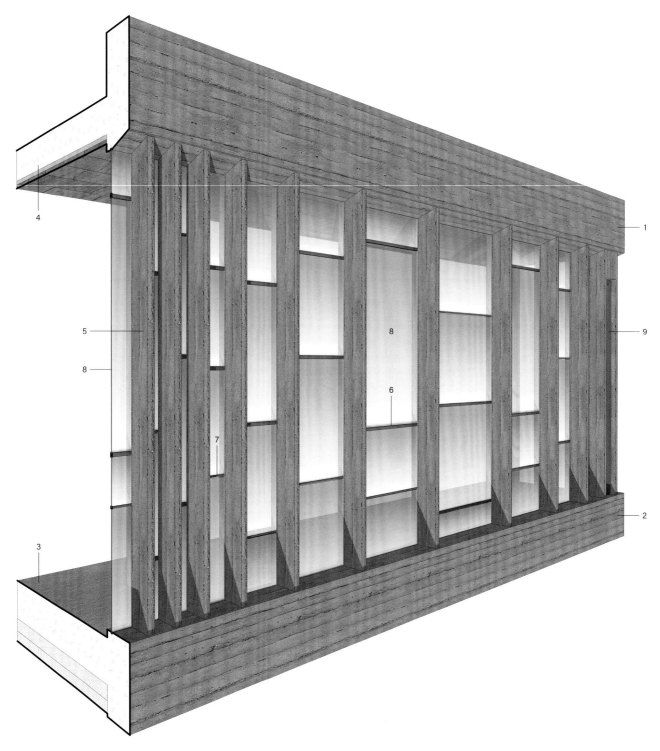

Typical section at *pans de verre ondulatoires* concrete frame window wall.

1 Reinforced-concrete roof beam
2 Reinforced-concrete floor beam
3 Reinforced-concrete floor
4 Reinforced-concrete slab
5 Reinforced-concrete mullion
6 *Pans de verre ondulatoires* metal frame
   (A: original; B: replacement)
7 *Pans de verre ondulatoires* non-metal frame
8 Single-pane clear glass
   (A: original; B: replacement)
9 Vertical pivoting operable aluminum shutter
10 Putty (A: original; B: replacement)

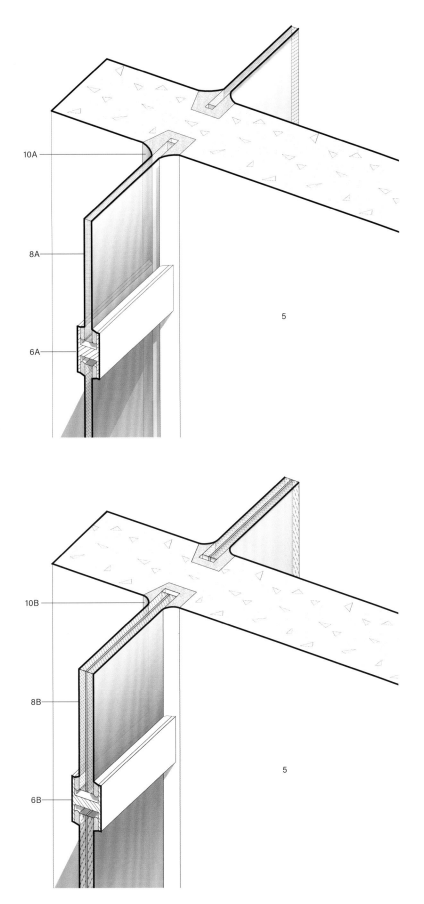

10A

8A

6A

5

10B

8B

6B

5

Typical before (top) and after (bottom) window wall details.

Interior view of the *pans de verre ondulatoires* from the ambulatory, 2014.

COMMENTS

The different interventions carried out on the *pans de verre ondulatoires* reflect the challenges of reconciling requirements for thermal performance and historic appearance when intervening on unique Modern buildings. At La Tourette, the replacement of the horizontal brass and silicone muntins and the retention of the new single-pane laminated glass ensure visual integrity and safety while reinstating the original appearance. However, this solution does not respond to contemporary requirements for energy conservation. The application of surface mounted solar protection films to prevent overheating of the interior spaces during the winter was discarded during the planning phase due to the adverse visual effect it would produce by distorting reflections on the exte-

Original vertical aluminum ventilation louver integrated into the concrete mullion shown in the closed, semi-open and open positions, 2014.

CREDITS

Convent of La Tourette (1960)
Éveux, France

ORIGINAL CONSTRUCTION

Architects
Le Corbusier with Iannis
Xenakis (Project Architects)
André Wogenscky with
Fernand Gardien (Executive
Architects, Site Supervisors)[11]

1960–1993 Interventions
Fernand Gardien (1960–1964)
Undocumented (1964–1979)
Jean-Gabriel Mortamet
(ACMH, 1982–1993)

2006–2013 INTERVENTIONS

Client/Owner
Convent of La Tourette

Sponsors
Velux Foundation
Foundation Spie Batignolles

Project Partners
Association of Le Corbusier
Sites; Utopies Réalisées;

Le Corbusier Foundation;
City of Firminy; Chapel of
Ronchamp

Architects
Repellin Larpin & Associés
Architectes
Didier Repellin (ACMH)

Collaborator
Pascal Duméril

Interior view of the reinforced-concrete mullions and replacement muntins in the refectory, 2014.

rior side of the glass. In her doctoral dissertation about the restoration of 20th-century glazed facades, Vanessa Fernandez commented that the intervention seemed anachronistic in that the facades were not adapted to meet contemporary standards which could have corrected the issue of the poor thermal performance of the glazed enclosure. When deciding between preserving the original appearance and complying with regulatory constraints, it seems like the decision-making process that led to the final solution would have benefitted from clear guidelines on how to mediate these conflicting interests.[10]

For the intervention on the glazed enclosures at La Tourette, reinstating and retaining the historic appearance was prioritized over achieving energy efficiency and thermal improvements. This was not the case at other contemporary buildings by Le Corbusier with similar *pans de verre ondulatoires* systems, such as the Maison du Brésil in Paris, where the original thin insulating glass panes were replaced with better-performing, slightly thicker ones. At the Convent, the reglazing could have been used as an opportunity to install solar films or insulating glazing to prevent the refectory and other spaces from overheating, and to improve overall interior comfort throughout. Instead, these issues were addressed by introducing mechanical ventilation, which increases the building service loads without improving the original building construction. The opportunity to sustainably improve the overall performance of the building complex of the Convent of La Tourette seems to have been missed.

NOTES

1 Father Pierre-Charles-Marie Couturier (1897–1954) had also previously commissioned Henri Matisse to decorate the Chapelle du Rosaire de Vence (Chapel of the Rosary) in the French Riviera.

2 Couvent de la Tourette, "History," accessed 25 February 2019 http://www.couventdelatourette.fr/the-building/history.html.

3 Cranston Jones, "Corbusier's Cloister," *Horizon 3*, no. 4 (March 1961): p. 38.

4 Fondation Le Corbusier, "Couvent Sainte-Marie de la Tourette," accessed 25 February 2019 http://www.fondationlecorbusier.fr/corbuweb/morpheus.aspx?sysId=13&IrisObjectId=4731&sysLanguage=en-en&itemPos=19&itemCount=79&sysParentName=&sysParentId=64.

5 Roberta Grignolo, "The Couvent de La Tourette from 1960 to the Present Day. Future Discernibility of Past Interventions" *DOCOMOMO Journal* 53 (February 2015): p. 66.

6 Grignolo, "The Couvent," p. 70, fig. 11.

7 Vanessa Fernandez, "Innover pour préserver. La restauration des façades vitrées du XXème siècle (1920–1970): De l'histoire des techniques à l'analyse des pratiques (Innovating to preserve. The restoration of the glazed facades of the 20th century (1920–1970): From the history of techniques to the analysis of practices)," (doctoral thesis, Université Paris, 2017), p. 181.

8 Grignolo, "The Couvent," pp. 67–68.

9 Ibid., p. 71.

10 Fernandez, "Innover pour préserver," p. 181.

11 Philippe Potié, *Le Corbusier. Le Couvent Sainte Marie de La Tourette/The Monastery of Sainte Marie de La Tourette* (Basel: Birkhäuser, 2001), p. 88; For more on the complex collaboration process, see: Karen Michels, *Der Sinn der Unordnung: Arbeitsformen im Atelier Le Corbusier* (Braunschweig/Wiesbaden: Vieweg, 1989).

Fagus Factory

Alfeld, Germany
Walter Gropius and Adolf Meyer, 1912/1925

Northeast facade of the original Fagus Factory in Alfeld, Germany, before completion of the existing addition, ca. 1911.

Southwest facade of the shop, new main (southwest) entrance and southeast facade after completion of the main building extension, ca. 1925.

The Fagus Factory in Alfeld was commissioned by Carl Benscheidt (1858–1947), who opened his shoe-last and cutting-die business in 1910 after leaving a competitor's facility located just across the train tracks from the site he had selected for his new venture. Benscheidt initially retained architect Eduard Werner (1847–1923) from Hanover for the project, who created the master plan and obtained local approvals to start construction. As work progressed though, Benscheidt grew dissatisfied with the appearance of the buildings from the adjacent train tracks. In 1911, he retained Walter Gropius (1883–1969), who redesigned the facades in collaboration with Adolf Meyer (1881–1929) to incorporate tetrapartite steel frame window walls and a bipartite bay at the northeast end. The original rectangular building, with the main (northeast) facade facing the railway line and including Gropius' and Meyer's steel frame exterior enclosure, was built between 1911 and 1912. The increase in production within a few years required a significant expansion of the main building, as well as of the storehouse, drying house and workroom. The result was a southwestward extension with a new southeast facade with a steel frame glazed enclosure similar to the original on the northeast facade. The new steel frame window wall wrapped around the corner, ending adjacent to the entrance at the new southwest facade.

Southeast corner of the factory with the steel frame window wall removed during the rehabilitation work, 1987.

Window wall reinstallation in progress at the southeast facade, where the steel frames were fitted with IGUs, 1989.

Window wall reinstallation in progress at the southeast corner, where the original single-glazing was reinstated, 1989.

Also designed by Gropius and Meyer, this is the existing front facade for which the building is most known since the addition was completed in 1925.[1]

Gropius and Meyer had worked together at Peter Behrens' office on the AEG Turbine Factory in Berlin, which was designed in 1908 and completed in 1909. With its robust textured concrete piers and slanted steel frame window walls between the semi-exposed steel columns, the AEG Turbine Factory had set a design precedent that informed Gropius' and Meyer's design for the Fagus Factory. Their solution for Carl Benscheidt's facility, however, included steel frame window walls with steel spandrels that projected off the masonry planes along the facades. Gropius later remarked on their design, saying that "the role of the walls becomes restricted to that of mere screens stretched between the upright columns of the framework to keep out rain, cold and noise."[2] Regarding the Fagus Factory, Nikolaus Pevsner wrote in *Pioneers of Modern Design* that, "for the first time a complete facade [was] conceived in glass," and that "thanks to the large expanses of clear glass, the usual hard separation of exterior and interior [was] annihilated."[3]

CONDITION PRIOR TO INTERVENTION

The factory complex has remained in continuous use since its original construction and has been fairly well maintained. Gropius' office was reportedly involved with the addition and several repairs until the mid 1920s. When Gropius visited in 1951, it was reportedly in excellent condition.[4] His direct involvement with the first addition and subsequent changes, and the fact that the com-

Northeast (right) and southeast (left) facades after rehabilitation and replacement, 2016.

plex escaped damage during both WWI and WWII and then was immediately listed as a protected heritage site in 1946 by the local authorities, helped to preserve it and retain its integrity. An economic downturn followed, however, and took a toll on the complex. By the 1970s, "large rust spots on the steel plates on the window frames" were noticeable on the exterior, along with other signs of decay.[5] Over time, both facades of the main building were treated with rust removal and protective coatings and the original steel frame window walls were replaced with similar systems manufactured by Hirsch and Fenestra.

A summary of the repairs undertaken over the years at the Fagus Factory prepared by Jürgen Götz, the engineer responsible for the renovation since 1982, included multiple details about the nature and condition of the original steel frame assembly. In it, Götz reported that the existing steel frame windows by Hirsch were corroded at the intersection of mullions and transoms.[6] The report also noted that the window wall headers had corroded and warped due to rust jacking, with some units corroded through,

Exterior detail of the single-glazed window wall at the northeast corner after rehabilitation, 2016.

Interior detail of the single-glazed window wall at the stair enclosure on the southeast corner after rehabilitation, 2016.

Southeast corner and facade after rehabilitation, 2016.

and that the steel channels along the slab edges were corroded as well. Götz found that the Fenestra products were drafty and the entire window wall assembly reportedly buckled under the high suction pressure generated by the exhaust system that was installed at the adjacent production hall to remove wood shavings. Conversely, when doors were slammed shut due to the strong draft, the window wall would buckle outwards. This resulted in considerable friction at the end of the transoms and mullions, which damaged the rust protection coating and further exacerbated steel corrosion. Prior to the renovation, the extent of corrosion was such that some windows were literally held together by putty. The frames around the pivoted steel windows sash rusted to the extent that they could no longer be opened or closed, which increased the draft even further.[7]

The report further notes that cast iron spandrel replacements were documented at the powerhouse, a change estimated to have taken place during WWI. The stair window wall had been replaced in the 1950s, apparently due to corrosion and deflection-related damage. Coating removal at a selected window wall location in 1994 revealed that various other alterations had taken place over the years, including selective replacement of steel members and spandrel steel sheets, and installation of additional rivets and bolts to reinforce the window wall around the operable

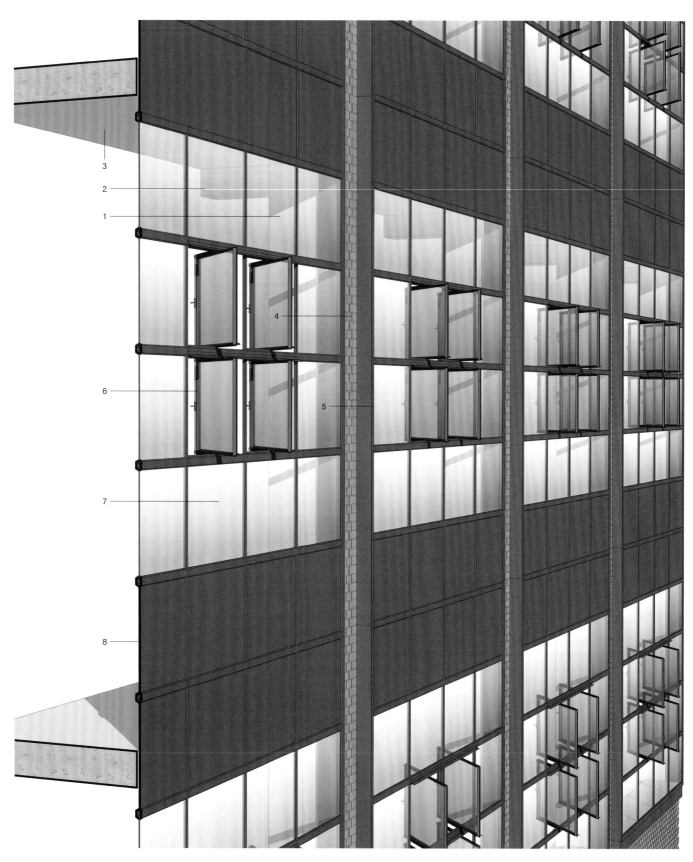

Typical section at original 1925 steel frame curtain wall.

1 Masonry-encased steel column
2 Concrete-encased steel beam
3 Reinforced-concrete slab
4 Brick masonry pier
5 Steel window wall perimeter frame
6 Steel frame vertical pivot window
   (A: 1912 original construction; B: 1925
   addition; C: ca. 1990 replacement)
7 Steel frame fixed window (A: 1912
   original construction; B: 1925 addition;
   C: ca. 1990 replacement)
8 Steel window wall spandrel
9 Original single-pane clear glass
10 Replacement IGU

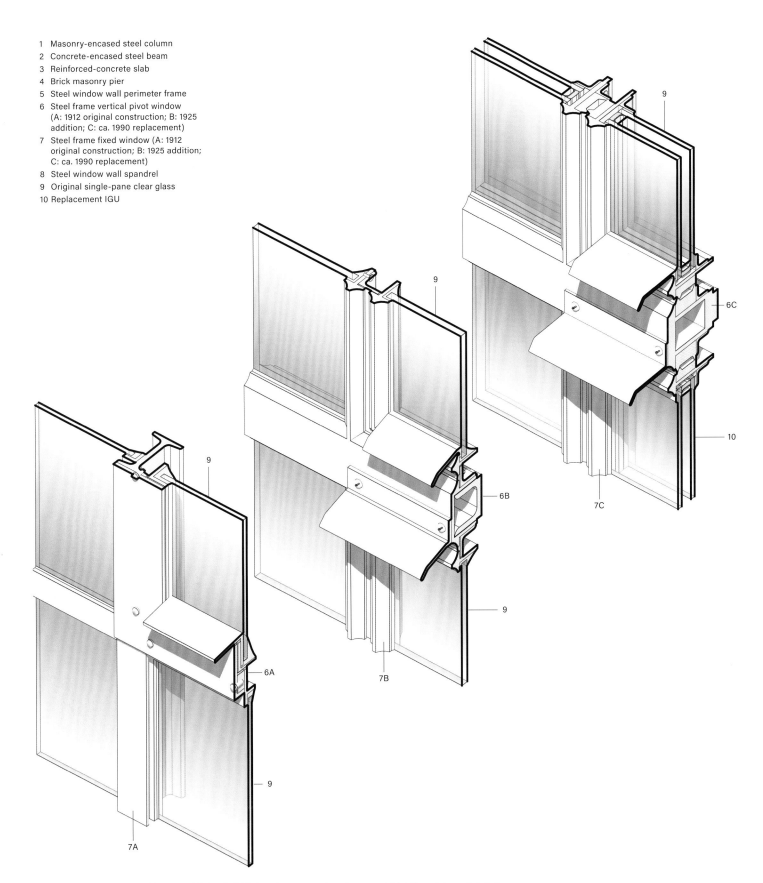

Typical original 1912 (left), 1925 addition (middle) and 1986–1990 replacement (right) curtain wall details.

units. The L-shaped supporting steel member along the bottom of the window wall had corroded and been replaced by a smaller profile. All of these undated changes were indicative that the facade and its original steel frame window wall required continued maintenance and multiple repairs shortly after construction.[8]

According to Götz, by the 1980s the ongoing deflections had resulted in an average glass-break rate of 40 panes per year. In addition to addressing the aforementioned physical decay and safety concerns, the building's occupants required improved environmental conditions in the office spaces, which for years had been drafty and subject to extreme seasonal heat and cold. As the raw material used to produce the boot last had changed by then from wood to plastics, there were no wood remains available to burn for heating and oil was expensive. Given that from the original inception continuity of use was one of the character-defining features of the complex, a change of use for the original office spaces was undesirable.[9]

INTERVENTION

The World Heritage site nomination for the complex indicates that the large window panels of the main building were irreparable as a result of corrosion and had to be replaced. For climatic reasons, the original single-pane glass on the facade of the workrooms had to be replaced with IGUs to improve thermal performance and retain the original use of these rooms. Restored window elements with the original single glazing and metal fittings were reinstalled in the hallways and in the southeast staircase.[10]

Design criteria were devised to guide the repair and replacement work at the Fagus Factory. The criteria were conceived to allow for the reuse of the steel frame members, and to maintain the subtle difference in aesthetics between the original northeast facade and the southeast facade of the addition. The intervention criteria were also meant to allow for visible changes only on the interior; to retain the number, location and operability of windows; to mimic the color and reflection of the original glass as much as possible; and to provide a protective coating matching the color of the original finish.[11]

The initial conservation plan in response to the intervention criteria, prepared by Jörn Behnsen in 1985, envisioned a complete window replacement that was rejected by the conservation authorities, who preferred retaining some of the existing components. An on-site peer review forum convened the same year was attended by various conservation experts and historians, who agreed on preserving and reinstalling wherever possible the original window components. The factory owners agreed to this initially for the

Detail of replacement windows with IGUs at the southeast facade, 2016.

Interior view looking up at the existing single-glazed steel frame window wall at the southeast corner staircase, 2016.

Detail of the replacement windows with IGUs installed at the southeast facade, 2016.

southeast staircase, and then later on for the conference rooms at the northeast corner, which had limited use.[12] The original steel frame single-glazed assemblies were thus retained at both end bays of the southeast facade, which includes part of the original facade and subsequent additions designed by Gropius and Meyer. Replacement IGUs were installed at the remaining window bays.

COMMENTS

One of the most striking features of the Fagus Factory, in addition to its transparent window wall and flat steel frames and spandrels, is the fact that it has been continuously used for the same purpose since its original construction. This low-rise, low-key, mid-sized, family-run manufacturing facility facing a rail yard in the midst of an industrial corridor seems to convey today the same unique sense of place that it did when it first opened more than a century ago. That sense of continuity and permanence is complemented by adaptations to contemporary requirements where needed, particularly at the window wall. The rehabilitation of the original single-glazed design at the end bays retains the original design intent where it has the most dramatic effect—at the corners. The choice to reinstate the original single-pane glazing at these corner locations and IGUs in between, while utilizing retractable awnings as a unified exterior window treatment throughout—

Exterior view of the existing single-glazed steel frame window wall at the southeast corner staircase, 2016.

Steel frame window wall at the southeast facade, 2016.

CREDITS

Fagus Factory (1912/1925)
Alfeld, Germany

ORIGINAL CONSTRUCTION

Architects
Walter Gropius and
Adolf Meyer

1982–1990 INTERVENTIONS

Owner
GreCon

Client
Institut für Denkmalpflege
Hanover

Architects
Architekturbüro Wilfried
Köhnemann (AWK)

Associate Architect
Jörn Behnsen

Structural Engineer
Jürgen Götz

Southeast and southwest facades, 2016.

all without altering the sightlines and configuration of the original window wall—is both practical and historically appropriate. It authentically presents the building's historic appearance and conveys its cultural significance as a relevant milestone in the origins and evolution of Modern architecture. This was accomplished while providing for environmental enhancements in the offices (where people are stationed for longer periods and user comfort is needed the most) and using a more forgiving, less restraining approach where people circulate or meet for short durations (such as at the southeast staircase and the conference room at the northeast corner). Furthermore, these subtle changes are embraced and acknowledged in the complex's interpretation plan, which clearly explains to visitors through audio guides in multiple languages how these changes contribute to enhancing the cultural significance of Gropius and Meyer's design for the Fagus Factory.

NOTES

1 Lower Saxony Ministry of Science and Culture and State Office for Monuments, *Nomination for Inscription on the UNESCO World Heritage List: The Fagus Factory in Alfeld*, September 2009, accessed 23 January 2019, http://whc.unesco.org/uploads/nominations/1368.pdf.

2 Walter Gropius, *The New Architecture and the Bauhaus*, (London: The MIT Press, 1937), pp. 22–23.

3 Nikolaus Pevsner, *Pioneers of Modern Design – From William Morris to Walter Gropius* (New York, The Museum of Modern Art, 1949), p. 214.

4 Jürgen Götz, "Maintaining Fagus," in *Fagus: Industrial Culture from Werkbund to Bauhaus*, ed. Annemarie Jaeggi, (New York: Princeton Architectural Press, 2000), p. 133.

5 Götz, "Maintaining Fagus," p. 133.

6 Ibid., p. 134.

7 Ibid., p. 137.

8 Ibid., p. 138.

9 Ibid., p. 139.

10 Wolfgang Kimpflinger, Wolfgang Neß, Reiner Zittlau, *Das Fagus-Werk in Alfeld als Weltkulturerbe der UNESCO: Dokumentation des Antragverfahrens* (Hanover: Lower Saxony Ministry of Science and Culture and State Office, 2011), p. 33.

11 Götz, "Maintaining Fagus," p. 140.

12 Ibid., p. 140.

Sanatorium Zonnestraal

Hilversum, Netherlands
Jan Duiker, Bernard Bijvoet and J. G. Wiebenga, 1928–1931

Driveway through the main building at the former Sanatorium Zonnestraal in Hilversum, Netherlands, ca. 1928.

Former driveway through the main building, mid-1980s.

Sanatorium Zonnestraal, literally meaning "Sunray Sanatorium" in Dutch, is located in the outskirts of Hilversum, roughly 18.5 miles (30km) from Amsterdam. This healthcare facility was designed in 1925–1926 by Dutch architect Jan Duiker (1890–1935) in collaboration with architect Bernard Bijvoet (1889–1979) and structural engineer Jan Gerko Wiebenga (1886–1974). Completed in 1931, the complex was built for the Diamond Workers Union of Amsterdam as a transitory facility, under the expectation that a cure for tuberculosis would be found within the next 30 to 50 years. Before effective cures using antibiotics were discovered in 1946, the rationale for building sanatoria in remote locations was that exposure to sun and fresh air was believed to be an effective treatment for tuberculosis. The remote locations also ensured that sick people would receive treatment away from the stigma associated with the infectious pulmonary disease.[1]

Former pavilion in ruins, 2002.

The design of the main building, adjacent wings and pavilions relies on a slender reinforced-concrete structure with tapered cantilevered beams around the perimeter, which help to maximize sun exposure and daylighting through the perimeter steel frame ribbon windows and window walls. As a way to minimize construction costs and with short-term durability in mind, the ribbon windows and window walls were specified without a galvanic coating. It was part of the patients' recuperation regime to protect the steel frames from corrosion by painting them regularly.[2] The main building, which was completed in 1928, was originally built with shallow 1" (25mm) deep steel frames and 5/32" (4mm) drawn glass (characterized by the wavy surface appearance left behind

Interior ground-floor corridor showing wall deterioration, steel window frame warping and broken glass, 1983.

Corridor facade detail showing corroded steel window frames, warped steel sashes and broken glass, 1983.

Corridor facade detail showing corroded steel window frames and detachment of the vertical cover plates, 1983.

by the rollers used during the manufacturing process). These steel frame ribbon windows were supported by a 3″ (76mm) wide flange steel beam (INP-8) mullion.

The Sanatorium Zonnestraal is one of the most significant examples of early Dutch Modernism, characterized by its simple unadorned facades that are delineated by reinforced-concrete aprons and cantilevered eaves, in turn defining modular openings glazed with large continuous steel frame ribbon windows and window walls. The complex is an embodiment of Duiker's ideas about the spiritual economy of efficient Modern design and the relationship between a building's form and its lifespan. The sanatorium is also is one of the most significant buildings produced by the Modern Movement in the Netherlands.

CONDITION PRIOR TO INTERVENTION

The 1928 windows had steel frames only on three sides (top, bottom and one side). These unconventional frames with shallow profiles allowed neither for sufficient installation tolerance, nor for expansion and compression of the steel. This led to excessive deflection at the transom bars, sash shifting and glass breakage at the 4′–11″ (1.5m) wide awning windows. To correct these deficiencies, the design of the steel frame glazed enclosures at Zonnestraal evolved as the complex grew. Subsequent pavilions completed in 1931 were built with standard four-sided steel frames comprised of wider and stiffer 1 1/2″ (38mm) profiles. These were casement ventilators instead of awnings, and the assemblies proved to be better suited for the height of the curtain walls at the stairs, as well as the individual size and overall length of the ribbon windows and perimeter window walls.[3]

The Sanatorium was in operation until 1950, closing four years after the discovery of an effective antibiotic treatment for tuberculosis, and much sooner than its original anticipated lifespan. In 1957, it was converted into a general hospital and various subsequent additions made the original main building and sunlit pavilions unrecognizable. When the hospital was vacated in 1993,

TIMELINE

1928	Main building completed
1931	Complex completion
1950	Sanatorium closes
1957	The former sanatorium is converted to a hospital
1982	Sanatorium abandoned; many windows broken, concrete fully exposed to the elements
1993	Hospital is vacated
1995	Master plan created by Hubert-Jan Henket Architects
1998	Buildings dismantled, demolition of many extensions to understand original configuration
2000	Permission granted to remove floor coverings and ceilings as well
2001– 2003	Interior and exterior of the main building restored
2002	First three workshops restored
2006	Fourth workshop renovated
2011	Included on the Netherlands' Tentative List for designation as a World Heritage Site by UNESCO's World Heritage Center

Replacement single-pane low-iron drawn glass at steel frame window wall, 2016.

the site was virtually abandoned. Severe deterioration followed. By the 2000s, most glass panes were missing and the steel frames were severely corroded, bent or split due to rust jacking. The buildings were overgrown with vegetation and all the interior and exterior finishes were soiled and damaged. The concrete structure had spalls, cracks and soiling from exposure to the elements, and it had collapsed at various locations.

INTERVENTION

In 1995, the rehabilitation of the complex as a healthcare center was outlined in a restoration master plan devised by Hubert-Jan Henket Architects. The restoration was later completed, supervised by Wessel de Jonge Architecten BNA B.V., in cooperation with the landscape architect Alle Hosper.[4] As per the master plan, the original 1928 workshops were renovated in 2000–2002 to serve as an obesity treatment clinic. The main building was rehabilitated in 2000–2003 to accommodate a sports injury rehabilitation clinic with a conference center on the upper floor of the main hall. The renovations included extensive facade repairs and reinstatement of original color at the replacement steel frames.

At the main building, part of the original steel frame ribbon windows could not be preserved due to their advanced deterioration and the fact that the double-glazing required to ensure a suitable indoor environment could not be fitted into the original 1″ (25mm) profiles. The new non-thermally broken replacement units are made of slightly thicker 1 1/4″ to 1 1/2″ (32–37mm) pro-

Corner detail of low-iron drawn glass at restored steel frame single-glazed window wall, 2016.

Close up of replacement non-thermally broken steel frame windows with IGUs made of slightly thicker profiles similar to the 1928 original, 2016.

files, similar to the improved original steel frames installed on the pavilions completed in 1931. Unlike the original 1928 units however, each fixed pane and operable sash has their own frame. To allow for minimum installation tolerances, the new back-to-back frames at the mullions are separated by a 1/8" (3mm) gap, permitting expansion of the steel work. This narrow joint is filled with sealant and tooled to express a subtle differentiation from the original assemblies.[5] The steel frames and glazing putty were painted with the original sky blue color, which was determined through historic paint analysis.

The replacement glass matched the appearance of the original drawn glass, with its slightly warped surface and resulting vertical distortions. It was produced in Lithuania, made with low-iron sand that is no longer commercially available in large quantities in Western Europe. The new panes not only matched the surface characteristics of the original drawn glass, but also its colorless visual properties. Single-glazed panes were installed along corridors and stairways, where lower building envelope performance was tolerable. At the workplaces, where a stricter building envelope performance is required, shallow 7/16" (11mm) IGUs were installed at the replacement steel frame. The makeup of the new IGUs, which were assembled by a Belgium fabricator, includes the same low-iron glass from Lithuania used for the single-glazed units on the outer pane and low-iron Starphire float glass by PPG imported from the US. The two panes are hermetically sealed with a neutral grey U-PVC spacer made in Italy.[6]

Typical light and visual distortions due to the roller wave effect of the replacement drawn glass at a steel frame parapet, 2016.

COMMENTS

The restoration of this significant Modern complex faced the challenge of finding a solution to the temporary nature of the original design. The approach selected for each building stemmed from careful considerations of the existing conditions, their degree of material integrity and their architectural significance. These decisions were guided by a set of intervention models developed for the project, which weighed architectural and historical perception (sensory perception and perception of original content) versus performance in use (functionality, comfort and management).[7] This exemplary methodological approach is one to be emulated on interventions in Modern buildings aimed at addressing functional obsolescence, particularly when decisions about authenticity and performance have to be made.

In terms of the glazed assemblies, the decision to forego the original configuration of the 1928 steel frame system at the main building and install a new system similar to the one at the 1931 pavilions struck the right balance between comfort and appearance. The use of low-iron drawn glass (an outdated technology that is now scarcely available) in single-glazed applications and IGUs demonstrates the project's commitment to both preserving

Replacement non-thermally broken steel frame windows with single-pane replacement glass along a circulation corridor, 2016.

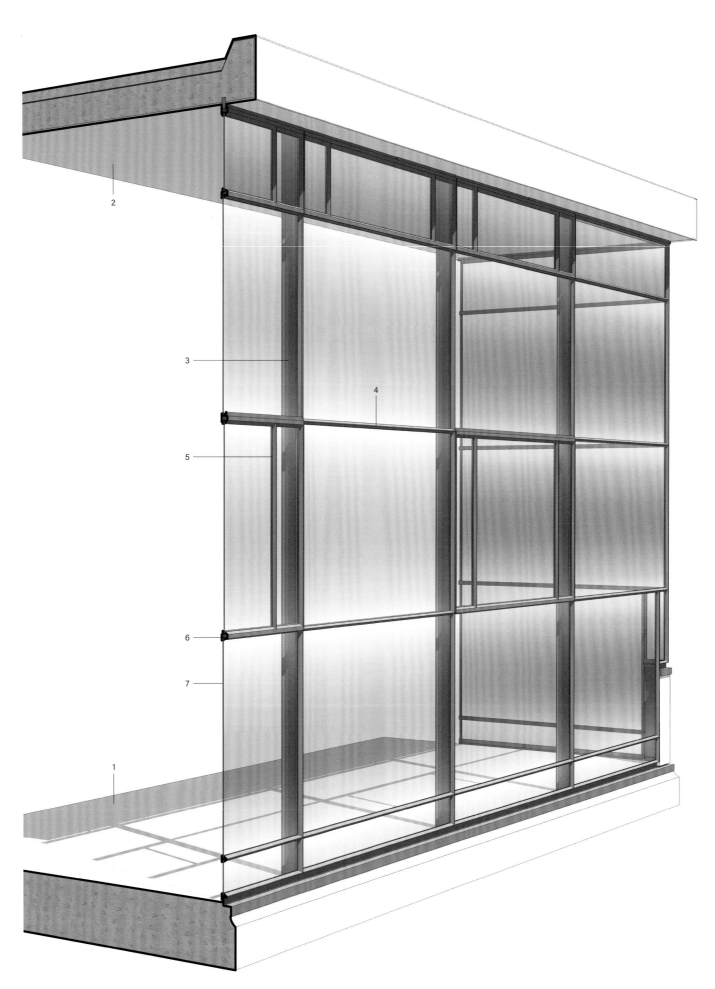

Typical section at original 1928 steel frame window wall.

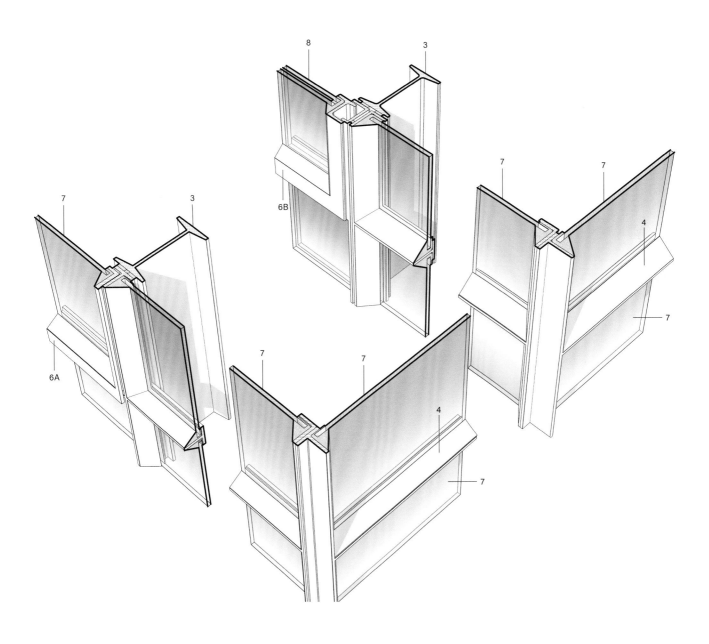

Typical before (left) and after (right) window wall details.

1 Reinforced-concrete floor slab
2 Reinforced-concrete roof slab
3 Steel window wall mullion
4 Steel window wall transom
5 Steel window wall muntin
6 Steel frame window (A: 1928 original construction; B: replacement similar to 1931 construction details)
7 Single-pane clear glass
8 Replacement clear IGU

Exterior view after the rehabilitation work and new all-glass exterior elevator addition, 2016.

the original design intent and meeting minimum requirements for performance durability and system integrity. It also enhances the visitor's experience, as the wavy vertical distortions of the large glazed enclosures are unique, creating memorable visual sensation that is unattainable at buildings with contemporary flat float glass.

The preservation approach of conserving the materiality and historic appearance of the exterior facades has complemented the new interior architectural programs at Zonnestraal. This is tangible where the interior circulation was placed along the exterior enclosures, thus allowing for as much single-glazing as possible. It is also apparent in that the extent of the replacement IGUs was limited to a window bay away from the corners, which allows the appropriate perception of weightlessness and visual transparency implicit in the original design intent. This solution emphasizes the

Restored non-thermally broken steel frame window wall with single-pane replacement glass at a stairway, 2016.

CREDITS

Sanatorium Zonnestraal
(1928–1931)
Hilversum, Netherlands

ORIGINAL CONSTRUCTION

Architects
Jan Duiker and Bernard Bijvoet

Structural Engineer
Jan Gerko Wiebenga

1998–2006 INTERVENTIONS

Client
Landgoed Zonnestraal BV

Architects
Hubert-Jan Henket,
Bierman Henket Architecten
Wessel de Jonge, Wessel de
Jonge Architecten

View of the restored steel frame single-glazed enclosures of the vertical circulation and adjacent pavilions, 2016.

original steel frame single-glazed corner details as valuable character-defining features that contribute to the significance of the complex within the Dutch and global contexts.

The interventions were not only restorative as new elements were discreetly introduced where needed. For instance, new window treatment, including interior blinds and exterior retractable awnings, were installed at south-facing facades. New interior fabric curtains were installed at the former dining and stage space at the second floor of the main building. These and other additions, such as the new glass-enclosed elevator shaft at the main building and the new steel frame connections between the workshops, are designed with a minimalist approach that is deferential to the original steel frame glazed enclosures. These design choices not only explore ways to express material authenticity, but also the elegance and refinement of the original design. The rebirth of Zonnestraal, and its careful balance between authenticity and functionality has won multiple accolades internationally. The intervention on this complex is also a unique example of how to successfully address functional obsolescence by implementing conservation and redevelopment plans that complement each other, are implemented in phases and assign suitable new programs to each space.

NOTES

1   Wessel de Jonge, "Zonnestraal: Restoration of a Transitory Architecture. Concept, Planning and Realisation in the Context of its Authenticity," in: *Proceedings: Seventh DOCOMOMO International Technology Seminar, Vyborg, Russia, September 18–19, 2003*, ed. O. Wedebrunn et al. (Copenhagen: DOCOMOMO, 2004), p. 2.
2   Ibid., p. 4.
3   Ibid., pp. 4–5.
4   Ibid., p. 6.
5   Ibid., p. 8.
6   Ibid., p. 9.
7   Paul Meurs and Marie-Thérèse van Thoor, eds. *Zonnestraal Sanatorium: The History and Restoration of a Modern Monument* (Rotterdam: NAi Publishers, 2011), p. 181.

City of Refuge

Paris, France
Le Corbusier and Pierre Jeanneret, 1933/1953

View of the original steel frame single-pane curtain wall facade of the City of Refuge in Paris, France, showing various combinations of glass (wired, clear and frosted) and window treatments (interior shades and curtains), ca. 1933.

View of the original steel frame single-pane curtain wall with the remedial sliding windows that were installed to improve user comfort, ca. 1935.

The City of Refuge is a homeless shelter in Paris designed for the Salvation Army by Swiss-French architects Le Corbusier (1887–1965) and Pierre Jeanneret (1896–1967). Located on a triangular city block, the main entrance into the complex is on Rue Cantagrel. From there, an elegant elevated portico and covered bridge lead into the lobby, which, in turn, leads to the main hall and the adjacent seven-story dormitory building with dining halls and other services on the lower floors. A two-story penthouse with smaller rooms for mothers with children is set back from the main facade of the dormitory building. Its zigzagging exterior masonry walls and south-facing steel frame window walls are visible from Rue Cantagrel. At the east end of the dormitory building, a narrow cascading facade on Rue Chevaleret provides both vehicular and pedestrian service access for the ground level and the garage in the cellar. Le Corbusier used different types of glass in different locations (wired glass at the spandrels, clear or frosted glass in the vision panes and patterned glass in the clerestory), experimenting with transparency and translucency to bring visual order to the large glass surface and to provide some privacy to the residents.[1]

For decades, architectural critics and historians described the design of the original facades as two parallel fixed single-glazed steel frame enclosures with a mechanically ventilated,

Original curtain wall as damaged during WWII, ca. 1944.

Replacement facade with concrete-framed grid of *brise soleil*, wood frame sliding windows and a new steel frame window wall at the penthouse, ca. 1953.

hermetically sealed interstitial cavity, coined the *mur neutralisant*, or "neutralizing wall."[2] However, research performed in preparation for the most recent building renovation determined that the original south and east facades consisted of a single steel frame curtain wall with no operable units.[3] The building mechanical system was supposed to provide purified forced air at a constant temperature to achieve an effect that Le Corbusier called "exact breathing" for a "factory for well-being."[4] Due to the Salvation Army's budgetary constraints though, a reduced version of the "exact breathing" system had been installed. It included low pressure steam heating loops, but no air conditioning.[5]

The work on each facade was awarded to two different contractors—on the narrow, east-facing Rue Chevaleret facade, Dubois & Lepeu (D&L) from Paris used standard U-, L-, T- and Z-shaped steel profiles. Along the Rue Cantagrel facade (south-facing), Menuiseries Métalliques Modernes (MMM) from Reims used stiffer and more expensive 1-1/4" (32mm) hot-rolled steel profiles for the large facade of the dormitory. Archival research concluded that the use of these dissimilar profile systems correlated with very different results. While the hot-rolled steel frames installed by MMM at the large south facade fulfilled the project's expectations regarding stability and weather tightness, the frames made out of standard profiles installed by D&L on the narrow east facade had to be reinforced afterwards.[6] The lack of operable windows on the south-facing glazed enclosure resulted in a greenhouse effect that was tolerable in the winter but unbearable during the summer. In response to complaints from the Salvation Army, wooden sliding interior shutters and curtains were installed on the interior side of the curtain wall to provide solar and visual protection. After a year of users' complaints—especially from the daycare staff and the single mothers caring for their babies—and per the advice of ventilation and air-conditioning specialists, the Salvation

TIMELINE

1929	First building designs
1930	Building permit for first three floors indicates operable windows were to be built
1931	Entire building permit is issued
1933	City of Refuge is inaugurated
1944	WWII bombing damages the glass facade
1948–1953	First building restoration campaign with new facade completed
1973–1977	Second restoration campaign completed
1975	Parts of the City of Refuge are listed and protected
1975	Hope Center extension completed
1986–1991	Third restoration campaign completed
2014	Completion of restoration phase 1 for the Hope Center
2015	Completion of restoration phase 2 or the City of Refuge, moderated by an Archaeological and Scientific Monitoring Committee

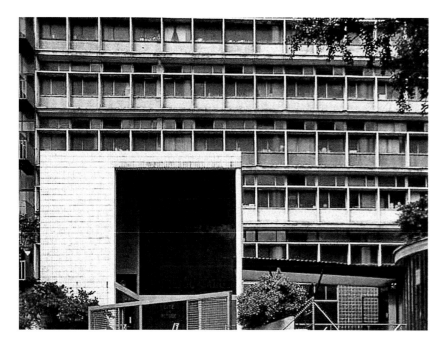

View showing replacement steel frame window walls at the penthouse and ground floor, and the concrete structure and *brise soleil* painted white, ca. 1975.

Fading paint and concrete soiling at the replacement 1953 facade, ca. 1975.

Army, in spite of Le Corbusier's objections, installed sliding windows on the last floor of the main south facade and at transoms in individual rooms.[7]

In 1944, a bomb blew up all the glass and brought to an end the short-lived existence of the original 10,764 sq. ft. (1,000m²) steel frame curtain wall that was the first of its size in Europe. The scope of the first post-war restoration from 1948–1953 included a redesigned facade among other work, and in 1950, Le Corbusier volunteered to design a replacement for the original curtain wall. His new design transformed most of the large south-facing facade along Rue Cantagrel into a concrete-framed grid of *brise soleil* (fixed sun-shading). However, with a depth of only 1'–6" (46cm), they were not optimal shading devices.[8] Behind this grid there was a conventional facade with wood frame windows laid on light masonry spandrels with fixed glazing in the center pane and a small operable transom. The smaller (east-facing) curtain wall on Rue Chevaleret was replaced with a similar facade.[9] Some of the original penthouse steel frame window walls and the ground floor storefront adjacent to the vehicular entrance remained in place. In 1952, disappointed by the client's choice of color for the exterior, Le Corbusier quit the job. More recently, the post-war 1953 replacement facades have been acknowledged in professional circles and by the public alike as being original parts of the building that continue to define the appearance and character of the City of Refuge today.[10] Their existence has also fueled the latest restoration discussions about which of the glazed systems (steel or wood frame) should be installed.

Entrance portico and reinforced-concrete facade, 2018.

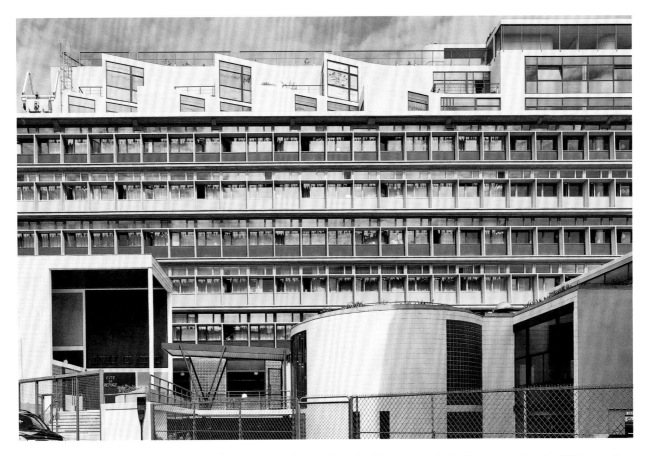

The penthouse and ground floor facades along Rue Cantagrel (south-facing) with replacement steel frames matching the 1933 original condition, 2018.

CONDITION PRIOR TO INTERVENTION

Following 1953, no major maintenance was undertaken at the building, and it eventually became dilapidated. A second restoration campaign from 1973–1977 repaired and replaced original elements and those of the post-war modification without any "consistent methodology or accurate historical and material knowledge."[11] The work included slight modifications of the wood frames to create operable windows.

The 1953 facades, the roof penthouse and main hall were listed as a National Historical Monument of France in 1975 in order to protect the building and its most character-defining features.[12] That same year, an adjoining building, the Hope Center, was built per designs by architects Georges Candilis and Philip Verrey to increase the capacity of City of Refuge. The two-story penthouse retained some of the original steel frame enclosures until 1975, when they were replaced with similar steel frame profiles. These and the 1975 steel frame system of the ground floor were retained until 2013.[13] The original low pressure steam loop heating system was also replaced by a conventional hot water system with radiators. The facade was further compromised by painting of the *brise soleil* and reconstruction of the entrance gateway.[14] The facade was painted according to a color scheme that was loosely based on Le Corbusier's 1952 project.

The 2015 restoration of the concrete *brise soleil* reinstated the painted spandrels and wood frame windows as designed by Le Corbusier in 1953, but with IGUs and retaining the operable transoms from the 1975 intervention, 2018.

View of the Rue Chevaleret (east) facade looking south after the 2015 intervention, 2018.

View of the penthouse replacement steel frame windows with clear and frosted glass and the restored reinforced-concrete facade below with replacement sliding wood windows, 2018.

The third intervention from 1986–1991 was aimed at meeting new building regulations, but was again executed without any historical survey. It led to major changes in substance and materiality—and notably was not fully coordinated with the Fondation Le Corbusier. In 1988, the wooden frames along Rue Chevaleret (east-facing) and Rue Cantagrel (south-facing) were replaced with aluminum sliding windows. The masonry spandrels were covered with aluminum panels and the windowsills with metal panning.[15]

Over the ensuing years, decaying materials and a series of localized alterations led again to a noticeable state of deterioration throughout the site. The draftiness of the facade had continued to worsen and the mechanical systems were in dire need of upgrading. Various shortsighted interventions had failed to holistically address the needs of the building and its occupants and ultimately created multiple adverse effects.

INTERVENTION

Finally, in 2007, the building's obsolete configuration and state of deterioration led the Salvation Army to launch a project for the rehabilitation of the entire complex. They joined with the social housing developer 3F in 2008, who had experience in project management and who contributed financial resources in the form of a long-term lease.[16] The architects in charge of the project brought in their own experience in social housing and historical monuments as well. Historical and conservation studies began in 2009, aiming to analyze the previous interventions and alterations. Eventually this process resulted in the implementation of an Archaeological and Scientific Monitoring Committee to moderate the discussion on conservation and restoration options. Launched in December 2011, the program was held in two phases. The first

The Rue Chevaleret (east) facade after the 2015 intervention, 2018.

Closer view of the Rue Chevaleret (east) facade after the 2015 intervention, 2018.

phase (completed in 2014), addressed the rehabilitation of the Hope Center.[17] The second project phase, which addressed the City of Refuge complex, was completed in 2015.

The penthouse and ground floor facades with steel frames from 1953 were restored to their original 1933 condition.[18] The main facade along Rue Cantagrel with the *brise soleil* was restored to the (protected) 1953 condition, but with wood frame windows that also included operable transoms like those installed during the 1975 alteration. The main objectives, however, were to maintain the original use according to current requirements and standards and to adapt the building to the City of Paris' climate plan by limiting the building's primary energy consumption to 80kW/m²/year.[19] As a result, all facade systems were retrofitted with new IGUs.[20] Although the facades were restored to their 1953 dimensions, the original bedrooms, whose sizing corresponded with the frame

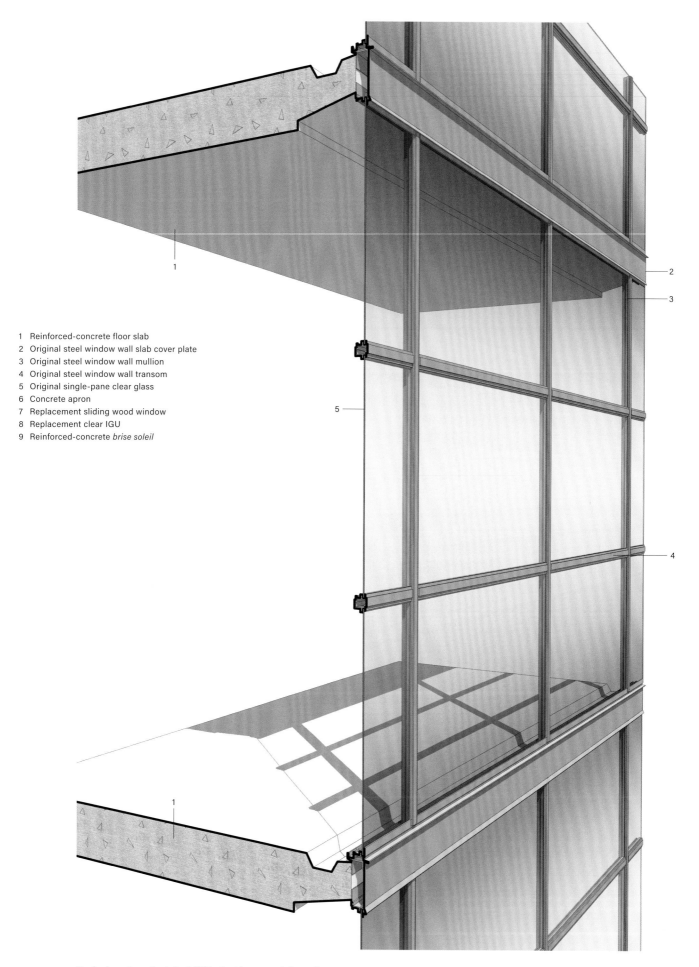

1   Reinforced-concrete floor slab
2   Original steel window wall slab cover plate
3   Original steel window wall mullion
4   Original steel window wall transom
5   Original single-pane clear glass
6   Concrete apron
7   Replacement sliding wood window
8   Replacement clear IGU
9   Reinforced-concrete *brise soleil*

Typical section at original 1933 steel frame curtain wall.

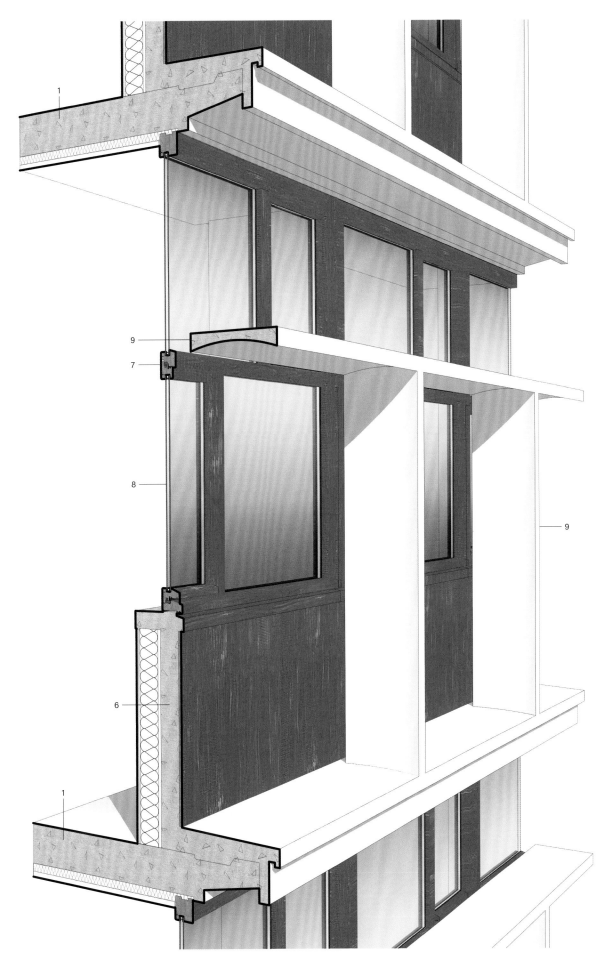

Typical section at 1953 reinforced-concrete facade with *brise soleils* and replacement wood windows with IGUs.

rhythm, were deemed too small and therefore unsuitable to address contemporary needs for single individuals, let alone couples or families.[21] This resulted in the removal of selected interior partitions to create new bedrooms approximately 8'-2" (2.5m) wide, instead of the original 6'-3" (1.9m) width. The original color finishes were reinstated on the outboard side of the 1953 concrete facade, which was repaired. The discussion about the original polychrome scheme of the facades was challenging due to the fact that only black and white photos existed. The final decisions were mainly based on exploratory probes, archival letters and narratives outlining the scope and materials included in contractor's estimates for previous interventions, all of which aimed to uncover Le Corbusier's original design intent for each significant intervention.[22]

COMMENTS

By embracing technological advances and foregoing ornamentation, Modern architecture sought new solutions to old needs. Experimentation and failure were therefore part of the quest for innovative solutions. This paradox is nowhere better illustrated than in the City of Refuge, with its failed, inoperable and non-functional original steel frame curtain walls and ensuing replacement facades with concrete *brise soleil*. The fact that Le Corbusier designed both facades raises the question of which one should have been reinstated. This question is better answered through the lens of authenticity—in other words, by determining whether values reside only in the building's materiality or also in the continuity of its use. The original facade design, damaged during WWII, had been altered to meet basic comfort and use requirements. The replacement facade, which to some extent allowed Le Corbusier to correct the failed sun protection of the original, had an even more significant effect—it allowed the building to retain its original use to date. That is why the Salvation Army's intervention focused less on preserving or reinstating the original design intent or the materiality of the facades and instead, understandably, more on adapting the building as needed to retain its original use at its original location.

According to the restoration project architect, "the goal wasn't to preserve Le Corbusier's ideas in amber, but to further the mission of the architecture by improving the building itself."[23] That meant retaining the original use by reinforcing its adaptability to contemporary requirements.[24] When assessed through this

Closer view of the Rue Cantagrel (south) facade and balconies after the 2015 intervention, 2018.

View of the Rue Cantagrel (south) facade and balconies after the 2015 intervention, 2018.

lens, the interventions seem sensitive and appropriate. However, in foregoing the original 1933 steel frame curtain wall altogether while restoring the 1953 concrete facade with wood windows and, at the same time, restoring the window walls at the penthouse facades with new steel frame assemblies (replacing the 1975 ones with the original 1933 configuration and retrofitting all with new IGUs), a new unintended paradox was created—one where the restored configurations had never coexisted before. According to one of the project team members, this was "not an ideal state in Viollet-le-Duc's sense of the term, but a state resulting from the desire to reconcile the restoration of heritage values with social issues, regulations and the client's budget."[25] To this end, the issue

The penthouse facade along Rue Cantagrel with replacement steel frame window walls matching the 1933 original configuration, 2018.

CREDITS

City of Refuge (1933/1953)
Paris, France

ORIGINAL CONSTRUCTION

Architects
Le Corbusier and Pierre Jeanneret, also responsible for the intervention from 1948–1953 (Pierre Jeanneret quit in 1950)

1973–1977 INTERVENTIONS

Architect (Restoration)
Philippe Verrey

**Architects
(Hope Center Extension):**
Philippe Verrey and Georges Candilis

1986–1991 INTERVENTIONS

Architect
Ljubomir Nikolic, ARENA

2007–2015 INTERVENTIONS

Architect
François Chatillon Architecte
Architecte en Chef des Monuments Historiques (ACMH)

Associate Architect
François Gruson, Opera Architectes

Engineering
COTEC

Architect and Researcher
Vanessa Fernandez (CSAS moderator)

Historian, Scientific Adviser
Vanessa Fernandez, Emmanuelle Gallo

Other Consultants
Fondation Le Corbusier

is then whether inaccuracies in the building's restored historic expression are detrimental to its authenticity and cultural significance or justifiable as a means of protecting its material fragility. The answer to that question will be a topic for years to come, one that will be assessed and influenced by the way that the building is presented to and interpreted by the general public. This seems only fitting for a design that was based on social and humanitarian ideas, intentions and values—one meant to express "humanity transmitted by an architectural object of value."[26]

NOTES

1   Emmanuelle Gallo and Vanessa Fernandez, "The glass facade and the heating system of the Salvation Army 'City of Refuge': from conception to restoration," In *DOCOMOMO Preservation Technology Dossier, no. 13: Perceived Technologies in the Modern Movement 1918–1975*, eds. Jos Tomlow, Alex Dill, Uta Pottgiesser (Zittau: DOCOMOMO ISC/Technology, 2014), p. 48.

2   Brian Brice Taylor, *Le Corbusier, the City of Refuge Paris 1929–1933* (Chicago: University of Chicago Press, 1987), p. 80.

3   Vanessa Fernandez and Emmanuelle Gallo, "A Factory for Well-being – Innovation in the Heating System and Curtain-Wall in Le Corbusier's Salvation Army City of Refuge, Paris 1933," in *Proceedings on the Eleventh DO-COMOMO International Conference, Mexico City, 20–25 August 2010*, (DOCOMOMO: 2010), p. 2.

4   According to Vanessa Fernandez (CSAS), who was part of the research team investigating the building prior to the last intervention, the *mur neutralisant* was intended for the Centrosoyous building in Moscow at the same time (1931–1932) and not for the City of Refuge in Paris. A mix of drawings of both projects led to the misinterpretation stated on various publications over the years. Rodriguez points out, however, that Le Corbusier did a laboratory experiment in 1935 to prove the efficiency of the double-glazed facade, probably aimed at solving the deficiencies identified at the City of Refuge curtain wall in Paris.

5   Fernandez and Gallo, "A Factory for Well-being," p. 3.

6   Ibid., p. 49.

7   Clarification provided to authors by Vanessa Fernandez.

8   Gallo and Fernandez, "The glass facade," p. 54.

9   Gilles Ragot, "Renovation and Restructuring the Cité de Refuge by Le Corbusier & Pierre Jeanneret–Preserving the Dual Functional and Architectural Identity of the Masterpiece," *DOCOMOMO Journal* 53, no. 2 (February 2015): p. 58.

10 Gallo and Fernandez, "The glass facade," p. 55.

11 Ragot, "Renovation and Restructuring," p. 58.

12 Ministère de la Culture, "Monuments historiques– Cité-refuge de l'Armée du Salut," last updated 13 October 2015, accessed 3 November 2018, http://www2.culture.gouv.fr/public/mistral/merimee_fr?ACTION=CHER-CHER&FIELD_1=REF&VALUE_1=PA00086591.

13 Gallo and Fernandez, "The glass facade," p. 54.

14 Ragot, "Renovation and Restructuring," p. 59.

15 Gallo and Fernandez, "The glass facade," p. 54.

16 Ragot, "Renovation and Restructuring," p. 57; Wolfgang Kabisch, "Betrifft. Die neue Cité de Refuge," *Bauwelt* 14 (2016): p. 9.

17 Architect François Chatillon. "Cité de Refuge: Restauration et restructuration de la Cité de Refuge pour l'Armée du Salut (hébergement social)," accessed 3 November 2018, http://architecte-chatillon.com/projets/cite-de-refuge/.

18 Ragot, "Renovation and Restructuring," p. 62.

19 Vanessa Fernandez, "Innover pour préserver. La restauration des façades vitrées du XXème siècle (1920–1970): De l'histoire des techniques à l'analyse des pratiques. (Innovating to preserve. The restoration of the glazed facades of the 20th century (1920–1970): From the history of techniques to the analysis of practices)," doctoral thesis, Université Paris, 2017, p. 183.

20 Gallo and Fernandez, "The glass facade," p. 55.

21 Architect François Chatillon, "Cité de Refuge: Restauration et restructuration de la Cité de Refuge pour l'Armée du Salut (hébergement social)," accessed 3 November 2018.

22 Ragot, "Renovation and Restructuring," p. 62.

23 Kelsey Campbell-Dollaghan, "The Radical Ideas Behind The Renovation of a Crumbling Le Corbusier Masterpiece," *Co Design*, last modified 4 May 2016, accessed 3 November 2018, https://www.fastcodesign.com/3058813/the-radical-ideas-behind-the-renovation-of-a-crumbling-le-corbusier-masterpiece.

24 François Chatillon, "Conservare è Moderno," *Domus*, no. 998 (January 2016): p. 39.

25 Ragot, "Renovation and Restructuring," p. 60.

26 Chatillon, "Conservare è Moderno," p. 39; Kabisch, "Die neue Cité de Refuge," p. 9.

Bauhaus Dessau

Dessau-Roßlau, Germany
Walter Gropius, 1926

View of the Bauhaus in Dessau-Roßlau, Germany, showing the workshop wing (in front), the school (at left in the background) and the residential studio wing (on the right), 1926.

Designed by German architect Walter Gropius (1883–1969), the Bauhaus Dessau is the pinnacle of pre-war Modern design in Europe. The Bauhaus School of Design and nearby Master Houses were the result of both the dissolution of the original school in Weimar and a desire by local politicians and art patrons to reconcile Dessau's industrial character with its cultural past.[1] The complex's carefully designed volumetric composition includes the four-story wing for workshops and studios, the technical school to the north, a two-story overhead bridge in between and a six-story residential studio wing to the east connected to the workshop wing by the low-rise cafeteria. Through its architecture, fixtures and furniture, the Bauhaus embodies the German concept of *Gesamtkunstwerk* ("synthesis of all arts"). The aesthetics of pragmatic functionality in the absence of ornamentation emerged from the Bauhaus to become the guiding paradigm of Modern architecture and design throughout most of the 20th century.

The original all-glass, steel frame curtain walls at the workshop and the ribbon windows throughout the building were manufactured by Nordische Eisen- und Drahtindustrie (Norddraht, or NORD-DRAHT).[2] They lacked any decoration and were limited to strictly functional components that provided structural stability, daylighting and natural ventilation. The curtain wall wraps the full length of the top three floors of the west, north and east facades of the workshop and residential studio wing. Its design is a refine-

The original steel frame curtain wall at the workshop wing, ca. 1927.

196

North wing, bridge and workshop wing shortly after completion in 1926.

The workshop wing with most of the original steel frame curtain wall removed after the damage suffered during WWII (note the brick masonry infill replacing the curtain wall), 1953.

ment and extension of the one that Gropius and his partner, Adolf Meyer (1881–1929), had designed for Karl Benscheidt, Jr. between 1911 and 1925 at the Fagus Factory in Alfeld (see p.166). This transparent enclosure clearly exposes the load-bearing reinforced-concrete and brick masonry structure behind, above and below the curtain wall while accentuating the latter as one of the building's main design features. At the bridge and school wing to the north, the steel frame single-glazed ribbon windows alternate with masonry aprons and dominate the facade composition. The interplay between ribbon windows and the painted stucco-finished masonry walls is at its highest at the taller residential wing, where small reinforced-concrete balconies cantilever away from the facade plane to comprise the much celebrated volumetric articulation for which the Bauhaus design is known.

CONDITION PRIOR TO INTERVENTION

After the Bauhaus was shut down in the fall of 1932, the complex had various uses including a school for training National Socialist district leaders.[3] It was targeted and bombed during a heavy air raid in 1945. Most of the original glass was blown up, and the original hot-rolled steel frame curtain wall and ribbon windows were severely damaged, with some areas completely detaching from the concrete slab structure. After the war, the facades of the workshop wing were significantly altered with brick masonry infill walls and wood frame windows, which remained in place until 1961. The curtain wall of the workshop wing and the stairwell windows were restored in 1976 with a non-thermally broken aluminum frame

TIMELINE

1925	Designed
1926	Completed
1945	School building damaged by bombing
1947	Partial restoration
1964	Included on the List of Monuments for the District of Halle
1965	Renovation of the facade and workshop wing
1974	Added to the former German Democratic Republic (GDR) List of Significant Monuments
1976	Comprehensive restoration
1994	Complex transferred to the Bauhaus Dessau Foundation
1996	Listed on the UNESCO World Culture and Nature Heritage List
1996– 2006	Restoration by Brambach und Ebert Architekten
2009– 2012	Window replacement and related interior work at north facade and residential wing

Front view of the workshop wing showing the existing replacement non-thermally broken aluminum frame curtain wall, 2016.

system, which has thicker mullions than the original steel frames. These replacement assemblies, although close to the original, "showed another construction principle, an altered function and a slightly different appearance."[4]

INTERVENTIONS

After the German reunification that followed the fall of the Berlin Wall in 1989, the Bauhaus Dessau Foundation was established as the complex's caretaker. A major restoration led by the Foundation was undertaken between 1996 and 2006. At that time, the 1976 replacement aluminum frame curtain wall was deemed to be in good condition and retained. Other windows, however, were reconstructed as per the original 1926 steel frame design. Several original 1926 windows were restored and retained. Other original 1926 windows removed during the 1976 intervention, which were thought to be lost, were found in an exterior greenhouse. They were documented and reinstalled in the building. This array of original 1926 steel frame windows, 1976 replacement aluminum frame curtain walls and 2000s steel frame window replications were all finished with the color of the original windows in an attempt to seamlessly preserve original character, relevant historical changes and contemporary replacements matching the original details.[5]

Three years later, in 2009, another intervention took place, this time addressing the workshop wing curtain walls, the ribbon

The existing replacement non-thermally broken aluminum frame curtain wall at the north facade of the workshop wing, 2016.

The existing replacement non–thermally broken aluminum frame curtain wall at the northwest corner of the workshop wing, 2016.

windows of the north wing and the windows and doors of the 28 studios in the residential wing. The main motivation was to create a more comfortable indoor environment at the Foundation offices (in the north wing) and at the residential studios to allow for year-round use and to reduce energy consumption and offset rising energy costs.[6] The replacement aluminum frame curtain wall at the workshop wing was again deemed in fair condition and retained as a "listed" significant building feature. Use of the interior spaces adjacent to the existing replacement curtain wall was reduced to seasonal functions during warm weather and the winter design temperature at the workshop wing (which is enclosed by large curtain walls on three facades) was set at 16°C. This lower target design temperature helped to reduce energy consumption in that wing by 50%.[7] Another measure implemented to compensate for the energy loss through the historic single-glazed curtain wall was the installation of a photovoltaic system on the north wing roof. Combined with the optimization of the heating and ventilation controls, this enabled the scope of the interventions to be reduced to a minimum.[8]

View looking up at the west facade of the workshop wing at the existing replacement non-thermally broken aluminum frame curtain wall, 2016.

The majority of the then-existing north wing ribbon windows, as well as the residential studio windows and doors were aluminum frame assemblies dating to the 1976 renovation. The original residential studio windows had a large casement, a small hopper and a small awning window. This arrangement was simplified in 1976 with fixed units and a large awning window. For the 2009 renovation, these non-original units were removed and replaced with steel frame assemblies matching the original 1926 sightlines and configuration of the operable windows. Instead of the original hot-rolled systems, new thermally broken steel frames were provided using a system based on flat laser-welded steel bars and U-profiles.[9] This manufactured system used a proprietary technology to create thermal separation using a fiber-reinforced plastic embedded within the width of the flat bars and between steel bars.

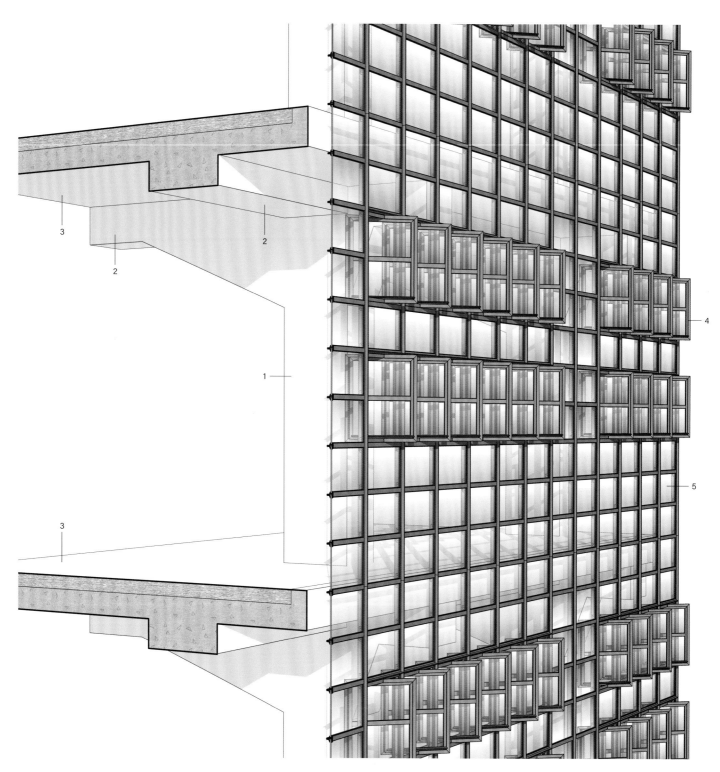

Typical section at original 1926 steel frame curtain wall at the workshop wing.

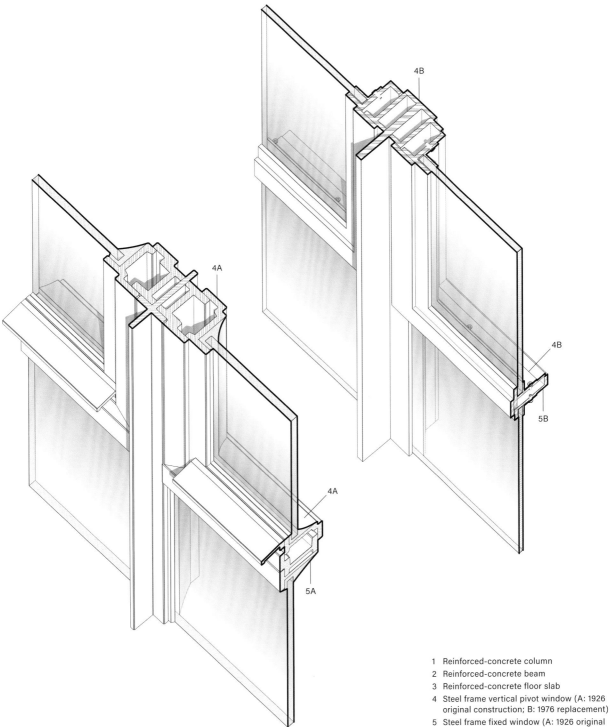

Typical original steel frame (left) and replacement aluminum frame (right) curtain wall details at the workshop wing.

1 Reinforced-concrete column
2 Reinforced-concrete beam
3 Reinforced-concrete floor slab
4 Steel frame vertical pivot window (A: 1926 original construction; B: 1976 replacement)
5 Steel frame fixed window (A: 1926 original construction; B: 1976 replacement)
6 Steel frame operable (awning, hopper and casement) ribbon window (A: 1926 original construction; C: 2015 replacement)
7 Steel frame fixed ribbon window (A: 1926 original construction; B: 1976 replacement; C: 2015 replacement)
8 Steel frame glazed door

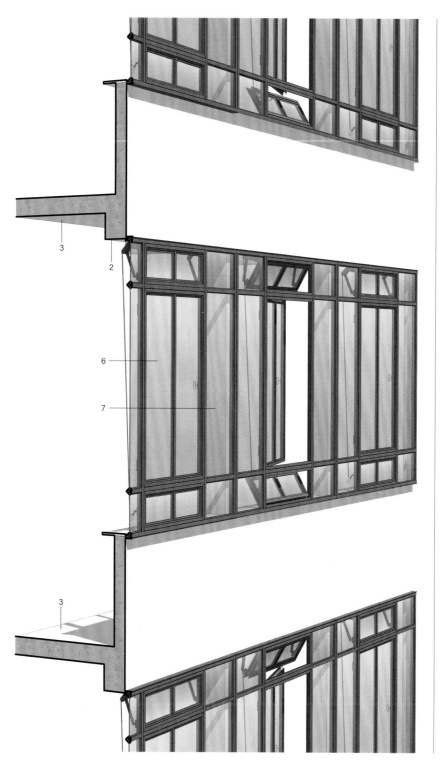

Typical section at original 1926 steel frame windows at the north facade of the north wing.

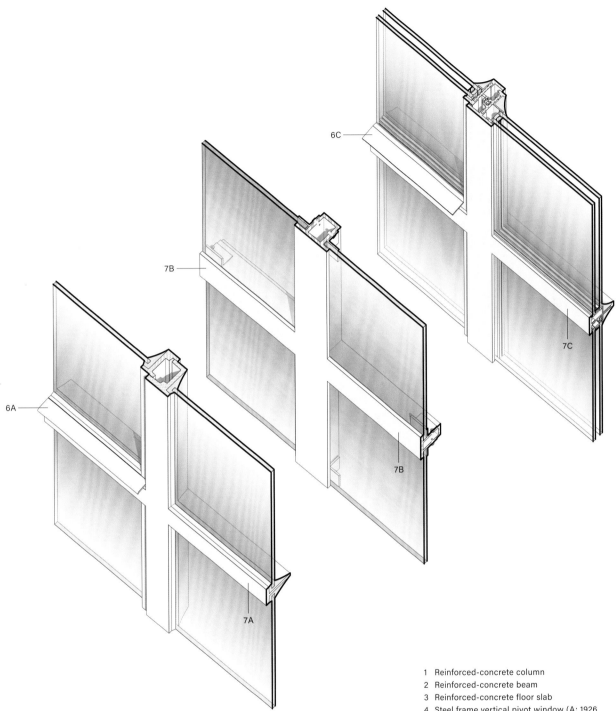

Typical original (left), 1976 replacement (middle) and 2015 replacement (right) ribbon window details at the north wing.

1   Reinforced-concrete column
2   Reinforced-concrete beam
3   Reinforced-concrete floor slab
4   Steel frame vertical pivot window (A: 1926 original construction; B: 1976 replacement)
5   Steel frame fixed window (A: 1926 original construction; B: 1976 replacement)
6   Steel frame operable (awning, hopper and casement) ribbon window (A: 1926 original construction; C: 2015 replacement)
7   Steel frame fixed ribbon window (A: 1926 original construction; B: 1976 replacement; C: 2015 replacement)
8   Steel frame glazed door

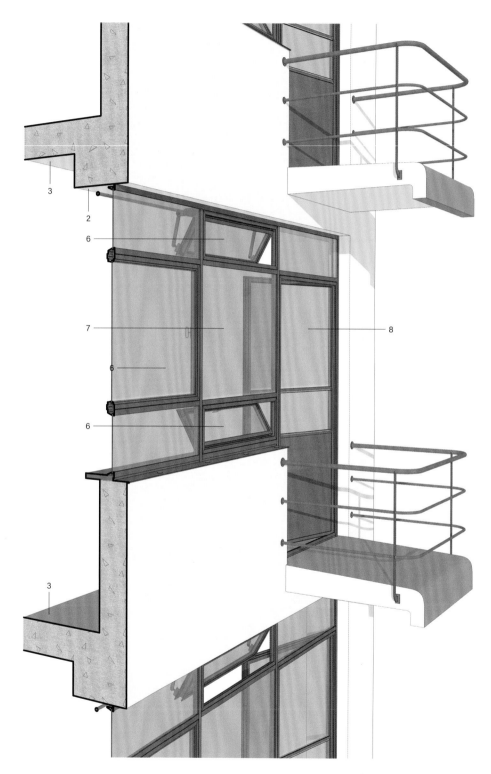

Typical section at original 1926 steel frame windows at the residential studio wing.

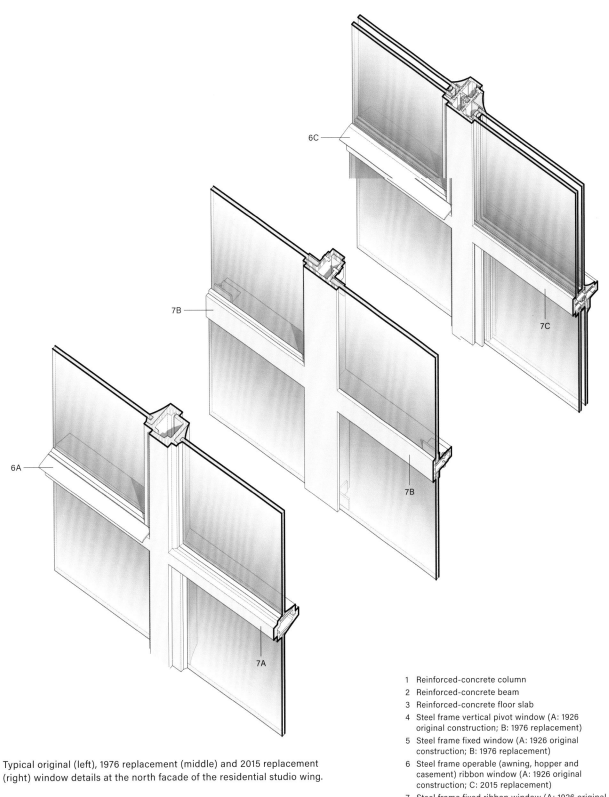

6C

7B

7C

6A

7B

7A

1 Reinforced-concrete column
2 Reinforced-concrete beam
3 Reinforced-concrete floor slab
4 Steel frame vertical pivot window (A: 1926
   original construction; B: 1976 replacement)
5 Steel frame fixed window (A: 1926 original
   construction; B: 1976 replacement)
6 Steel frame operable (awning, hopper and
   casement) ribbon window (A: 1926 original
   construction; C: 2015 replacement)
7 Steel frame fixed ribbon window (A: 1926 original
   construction; B: 1976 replacement; C: 2015
   replacement)
8 Steel frame glazed door

Typical original (left), 1976 replacement (middle) and 2015 replacement
(right) window details at the north facade of the residential studio wing.

Interior view of the existing replacement non-thermally broken aluminum frame curtain wall at the staircase, 2016.

Interior view of the existing non-thermally broken steel frame basement windows, 2016.

Interior view of the replacement non-thermally broken aluminum frame curtain wall at the workshop wing, 2016.

This renovation allowed for another rectifying change and improvement regarding the glass. The original 1/4" (6mm) plate glass had been made through an elaborate and expensive polishing process that resulted in mirror-like ultra-clear glass with very few visual distortions. This fabrication system is no longer available today. The low reflection effect from the original can be achieved though through today's standard float glass, which has the identical quality in flatness. Replacement single-pane glass, however, would have made a poor contribution to the thermal performance of the new units, particularly taking into account the energy analysis for the complex which determined that the largest energy losses were related to the glazed building envelope. After evaluating the visual properties of several IGUs with low-iron glass panes and krypton infill, a consensus was reached to use low-iron glass but forego a low-emissivity (low-e) coating on the interior of the IGU. The coating would have improved the heat re-

Sill detail of ribbon windows with a combination of original steel frame and replacement non-thermally broken aluminum frames at the south facade of the north wing, 2016.

Head detail of ribbon windows with a combination of original steel frame and replacement non-thermally broken aluminum frames at the south facade of the north wing, 2016.

tention properties of the IGUs but would have lent an undesirable color graduation to the otherwise clear glass. Instead, to enhance thermal performance, the replacement windows were supplemented by the use of insulating plaster on the interior and low-voltage heat tracing along the existing non-thermally broken steel angle supporting the ribbon windows in the north wing, neither of which are visually perceptible.[10] However, it was anticipated that the changes would not be sufficient to prevent interior condensation during the coldest winter days on their own, so they were supplemented by initiatives to raise user awareness. This included providing information about the limitations of the thermal capacity of the historic enclosure and instructions for ventilation and localized electronic moisture monitoring with optical signals calling for fresh air to minimize indoor moisture when required.[11]

COMMENTS

The renovation of the glazed enclosures of the Bauhaus holds unique relevance within the field of conservation of Modern architecture. Faced with the combination of original and reconstructed assemblies made of different materials and dating to various construction periods, the latest intervention approach did not focus on material authenticity within the glazed enclosure, but on retention of the historic image and character. For instance, the retention of the replacement 1976 aluminum frame curtain wall at the workshop is a unique example of intervention restraint where the impetus to remove all non-original material and replace it with new state-of-the-art high-performance systems was counterbalanced by an unequivocal willingness to assign value to prominent contributions accrued over time. Thus, the non-original aluminum frames from the curtain wall at the workshop, which depart from the original details but mimic the original sightlines, have been retained and only the glass has been replaced with single-pane low-iron glass to reinstate translucency. On this part of the intervention, reestablishing the historic appearance was appropriately deemed more important than enhancing energy efficiency.

Existing replacement thermally broken steel frame ribbon windows at the north facade of the north wing, 2016.

Conversely, a different but similarly appropriate approach was pursued at other locations like the north facade of the school and the east facade of the residential wing, where the 1976 replacement aluminum frame glazed assemblies were removed and replaced with a thermally broken steel system. At these locations, where the glazed assemblies are smaller and less prominent than at the workshop, comfort and energy savings goals were prioritized. The restrained approach balanced performance and use

CREDITS

Bauhaus Dessau (1926)
Dessau-Roßlau, Germany

ORIGINAL CONSTRUCTION

Architect
Walter Gropius

1996–2006 INTERVENTIONS

Architects
Pfister Schiess Tropeano &
Partner Architekten AG, in

association with Arge Bauhaus,
Brambach + Ebert Architekten,
Halle/Saale
Pfister Schiess Tropeano &
Partner Architekten AG Zürich

Research
Monika Markgraf and Johannes
Bausch (the Bauhaus Dessau
Foundation)

2009–2012 INTERVENTIONS

Architecture and Planning
Brenne Architekten GmbH

Construction Management
Brenne Architekten GmbH

Mechanical Engineering
GfE Gesellschaft für
Energieeffizienz mbH

Electric Design
Bauer & Zuber Elektroplanung

Climate Engineering
Transsolar Energietechnik
GmbH

Window Manufacturer
MHB from Herveld,
Netherlands

Existing replacement thermally broken steel frame windows and doors
at the residential studio wing, 2016.

(similar to the approach at the workshop), but also implemented a
suitable combination of interventions at adjacent locations, such
as using an insulating plaster finish on the masonry aprons and
implementing supplemental low-energy active solutions like localized heat tracing and use-based adjustments to building systems when required. At the Bauhaus, continuity of use and functionality are embraced as conservation tools. The interventions
have successfully prioritized reinstating the technical and functional aspects of significant building features over adherence to
original materials and systems. This approach is a result of the
continued reinterpretation of the site as "functional heritage".[12]

The effects of the plurality of construction dates throughout
the complex are subtle, but still obvious to an educated eye. Yet
this does not distract from perceiving a coherent image and experiencing the building's significance within the origins and evolution of the Modern Movement. No attempt is made to adhere to
one specific period over another, nor to differentiate the products
from one period over another. All periods are deemed significant
and valued, as a condition for conservation. Continuity and
change, past and present, are embraced as avenues for character
retention. Functionality within an unadorned aesthetic—a core
concept within the Modern vision that the building represents
more than any other exponent from the 20th century—is what is
protected at the Bauhaus as an original design concept that em-
braces past and existing changes.

NOTES

1   Reginald Isaacs, *Gropius: An Illustrated
    Biography of the Creator of the Bauhaus*
    (Boston: Bulfinch Press, 1991, 1st English-
    language ed.), p. 119.

2   Christopher H. Johnson, "Steel Window
    Reconstruction at the Bauhaus Dessau –
    A Recent Case Study in the Practice of
    Renewing Modernism," Brandenburg Univer-
    sity of Technology, p. 7, accessed 23 January,
    2019, https://www.academia.edu/4187937/
    Steel_Window_Reconstruction_at_the_
    Bauhaus_Dessau.

3   Bauhaus Dessau. "The Bauhaus Building –
    history of use," accessed 11 November 2018,
    https://www.bauhaus-dessau.de/en/history/
    bauhaus-dessau/the-bauhaus-dessau.html.

4   Monika Markgraf, "Conservation and
    Preservation of the Bauhaus Building in
    Dessau," in *Heritage at Risk – Special Edition
    2006: The Soviet Heritage and Europe
    Modernism*, Chapter IV. World Heritage Sites of
    the 20th Century – German Case Studies, eds.
    Jörg Haspel, et al. (Berlin: Hendrik Bäßler
    Verlag, 2007), p. 110.

5   Ibid.

6   Winfried Brenne and Ulrich Nickman, "Neue
    Fenster fürs Bauhaus, DE-Dessau," *Architektur
    + Technik* (August 2012), p. 97, accessed 11 May
    2019, http://www.architektur-technik.ch/Web/
    internetAxT.nsf/0/5845EF24E2DDBD5EC1257-
    A83004EC0CC/$file/096-100%20Bauhaus-
    Fenster.pdf?OpenElement.

7   Ibid., p. 98.

8   Johnson, "Steel Window Reconstruction," p. 10.

9   Brenne and Ulrich, "Neue Fenster fürs
    Bauhaus, DE-Dessau," p. 100.

10 Ibid.

11 Johnson, "Steel Window Reconstruction," p. 15.

12 Ibid., p. 3.

Solomon R. Guggenheim Museum

New York, New York, USA
Frank Lloyd Wright, 1959

The Monitor building north of the Rotunda at the Solomon R. Guggenheim Museum in New York, New York, ca. mid-1960s.
© Ezra Stoller/Esto

The Solomon R. Guggenheim Museum in New York City, designed by American architect Frank Lloyd Wright (1867–1959) and completed in 1959, is a combination of two well-defined structures known as the Monitor and the Rotunda, which complement each other both functionally and aesthetically. The Rotunda is the spiral exhibition space for which the Guggenheim—if not Wright himself—is better known worldwide. Wright designed the Monitor as independent levels connected to the Rotunda ramps, with most of the two upper floors enclosed by polygonal floor-to-ceiling, window walls leading to perimeter balconies. The original glazed enclosures were manufactured by the US branch of Henry Hope & Sons from the UK. They were comprised of a single-glazed uninsulated galvanized steel frame system with a painted metallic finish.

Over the years, the administrative functions Wright envisioned for the Monitor were relocated elsewhere into newly created spaces; first in 1968, within a concrete-framed addition designed by William Wesley Peters of Taliesin Associates, and then again after a major renovation and expansion in 1992. This expansion included the completion of an eight-story steel-framed addition designed by Gwathmey Siegel & Associates. Since 1976, following the removal of the interior partitions and completion of related renovation work, the Monitor has been known as the Thannhauser Gallery.

The Guggenheim Museum, notwithstanding the alterations and additions that have taken place, is a prime example of Wright's later work, manifesting his theory of "organic architecture" in the unity of building method, appearance and use. Nowhere in the Guggenheim is Wright's formulation of organic architecture more evident than along the window walls at the former Monitor, where the exterior glazed enclosures embrace rather than separate the indoor and outdoor environments that they are meant to demarcate.

CONDITIONS PRIOR TO INTERVENTION

Field evidence indicated that the original galvanized steel framing of the window walls was in good condition, showing no sign of deterioration. By traditional preservation standards, they would have been deemed sound and the original historic fabric retained. However, during the winter, water vapor from warm, humidified indoor air condensed on the colder surfaces of the uninsulated glass panes, steel frames and mullions. Under extreme winter conditions, the Museum's indoor environmental targets 20–22°C dry bulb, 50–55% relative humidity) allowed this condensation to freeze and form icicles along the steel frames.[1]

Similar condensation occurred during the summer, although to a lesser extent, as outdoor atmospheric water vapor condensed on the colder exterior glass surfaces of the Thannhauser Gallery. This seasonal condensation not only impaired indoor/outdoor visibility, but led to energy loss and increased demand for air conditioning, imposing greater heating and dehumidification loads on the building's mechanical equipment. It compromised the stability of the Museum's indoor environment and rendered the Thannhauser Gallery unfit for exhibitions during cold seasons.[2]

TIMELINE

1943–1956	Design period
1957	Window wall installation
1959	Building completion
1968	Four-story concrete-framed museum tower addition completed
1990	Building is designated as New York City Landmark and the Rotunda is designated as a New York City Interior Landmark by the Landmarks Preservation Commission
1990	Construction begins to renovate and expand the museum interior
1992	Museum interior renovation and eight-story steel frame tower addition by Gwathmey Siegel & Associates are completed
2004	Listed in the National Registry of Historic Places; project for the exterior restoration of the Rotunda and Monitor begins.
2005	Refurbishment and replacement options for the glazed enclosures conceived and evaluated by WASA/ Studio A
2008	Window wall removal and Replacement work completed
2008	Building is designated as a National Historic Landmark by the US Secretary of the Interior
2019	Designated a UNESCO World Heritage Site (with seven other buildings designed by Frank Lloyd Wright)

Original steel frame single-glazed windows at the third floor during construction, 1957.

Original steel frame single-glazed window wall at the fourth floor during construction, 1957.

INTERVENTION

More than a dozen options were considered to address the undesirable effects of seasonal condensation at the Guggenheim Museum. These options followed two main approaches—refurbishment or replacement. Energy models were devised for each of the proposed refurbishment and replacement systems, using software by the Windows & Daylighting Group from Lawrence Berkeley National Laboratory. The pros and cons of each approach were considered. Unsurprisingly, the results of the energy models showed lower performance for refurbishment options versus replacement. The study also concluded that besides limited performance enhancement, the solutions derived from the refurbishment approach had other relevant implications. For instance, this approach would have significantly modified the appearance of the historic metalwork and, to some extent, would have been injurious to sound original fabric. Options to retain the well-preserved steel frames were not warrantable and their performance had limited predictability.[3]

Despite the good physical condition of the steel framed window walls, their poor thermal performance and lack of airtightness made replacement the more appropriate upgrade option. For the window walls, enabling the year-round visibility implicit in the original design intent and upgrading the environmental performance took precedence over retaining original historic fabric in good physical condition. Accordingly, an innovative thermally broken steel frame system, matching the appearance and sightlines of the original window wall, was designed by WASA/Studio A in collaboration with William B. Rose & Associates. The replacement system included IGUs fitted with warm-edge technology along the spacers, 90% argon infill and a low-e coating on glass surface #3. Construction details were prepared by selected manufacturers to build the thermally broken systems and perform a comparative

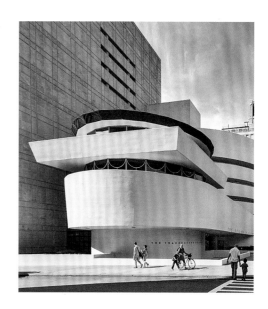

Exterior view after the rehabilitation work and window wall replacement, 2008.

Interior condensation at the original third-floor steel frame single-glazed window wall, 2005.

Interior condensation at the original fourth-floor steel frame single-glazed window wall, 2005.

Detail of original non-thermally broken steel frame window wall, 2005.

Detail of replacement thermally broken steel frame window wall, 2008.

testing of such systems. The successful replacement option, which retains the original appearance and intent, was an unprecedented custom thermally broken steel frame system matching the exterior appearance of the original glazing. It was also the first thermally broken steel frame system installed in the US.

The removal of the original poorly performing galvanized steel frame single-glazed window walls at the Guggenheim and subsequent replacement with custom thermally broken steel frame system with IGUs successfully eliminated the seasonal condensation that had limited the use of the former Monitor since the original construction. The condensation having been addressed, the positive effect of a more efficient glazed enclosure enabled the mechanical equipment serving this portion of the building to be downsized.[4] The replacement intervention also met the historic preservation goal of minimizing its impact on the material nature and overall appearance of the glazed enclosures as character-defining features. Only close inspection by a trained eye might notice that they are not original. The well-disguised grey warm-edge spacers match the color of the grey metallic steel paint but, to the careful observer, the IGUs discreetly reveal themselves as contemporary features. More importantly, year-round use is now programmed for the renovated spaces, which are used primarily as a cafe from where visitors can enjoy Central Park that, according to Wright, "ensures light, fresh air and advantages in every way but one and that is congestion."[5]

COMMENTS

In 2005, no thermally broken steel frame glazed assemblies were commercially available in the US. When asked why not, the local representative of a large, well-established steel frame window manufacturer responded that no one was asking for them, so there was no need to produce them. Ten years after the research

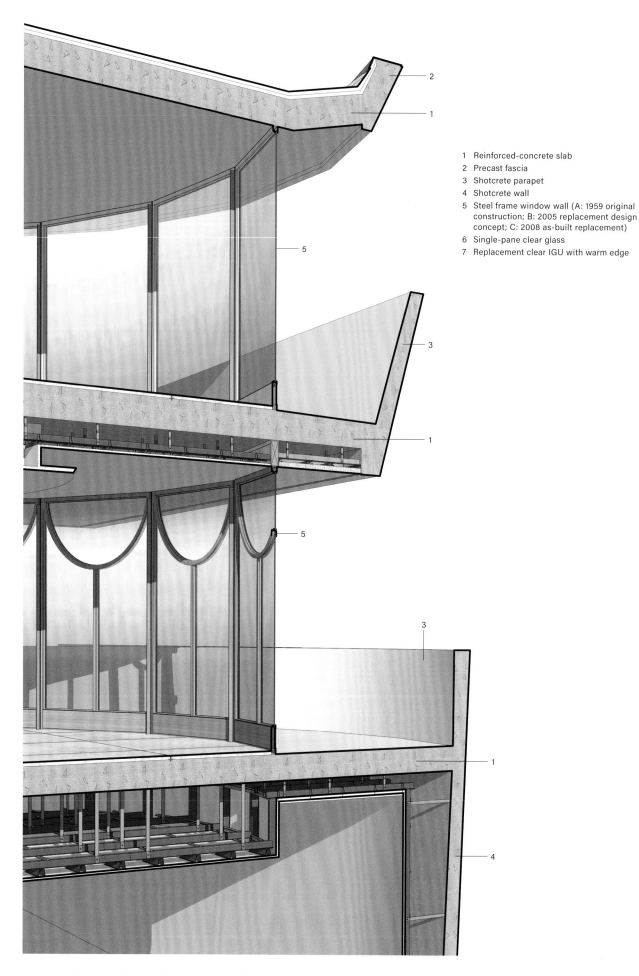

1 Reinforced-concrete slab
2 Precast fascia
3 Shotcrete parapet
4 Shotcrete wall
5 Steel frame window wall (A: 1959 original construction; B: 2005 replacement design concept; C: 2008 as-built replacement)
6 Single-pane clear glass
7 Replacement clear IGU with warm edge

Typical section at existing steel frame window walls before the replacement work.

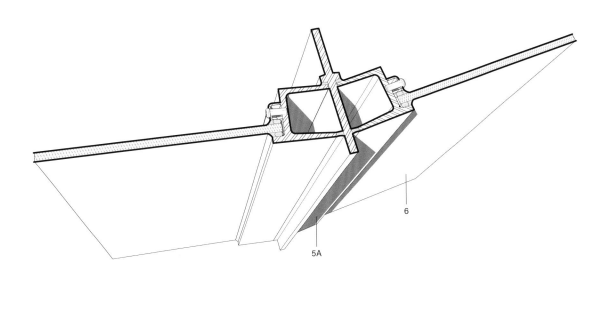

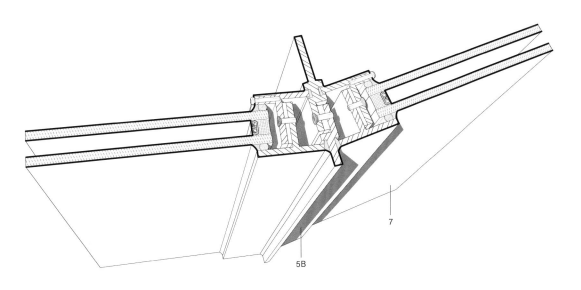

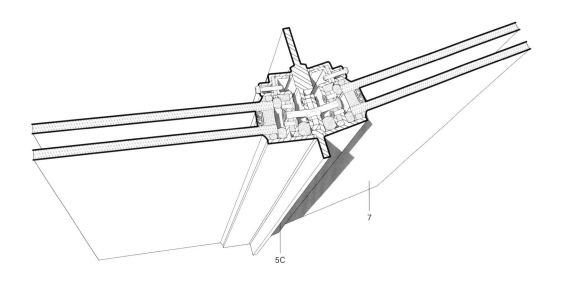

Typical original (top), proposed (middle) and as-built replacement (bottom) window wall details.

Interior view of the third-floor replacement thermally broken steel frame window wall after the space was renovated as a café, 2014.

Interior view of the fourth-floor replacement thermally broken steel frame window wall, 2012.

that led to the removal and replacement of the original non-thermally broken single-glazed window wall at the Guggenheim, over a dozen manufacturers were offering thermally broken steel frame systems in the US and Europe. Thermally broken assemblies, which for decades were only available in aluminum systems of dimensions larger than steel sections, are now available in other metals such as bronze, stainless steel, weathering steel, bimetals and metal-wood combinations. Available products include both cold-formed and hot-rolled assemblies with machined grooves that accommodate high performance isobars connected to the steel through proprietary systems.

CREDITS

Solomon R. Guggenheim
Museum (1959)
New York, New York

ORIGINAL CONSTRUCTION

Architect
Frank Lloyd Wright

2005–2008 INTERVENTION

Owner/Client
Solomon R. Guggenheim
Museum and Foundation

Project Director
Paratus Group

Architects
Wank Adams Slavin Associates
LLP (WASA/Studio A)

Structural Engineers
Robert Silman Associates,
PC (Silman)

Architectural Conservator
Integrated Conservation
Resources (ICR)

Exterior Envelope Consultant
William B. Rose & Associates

M/E/P Engineers
Atkinson Koven Feinberg
Engineers, LLP (AKF)

**Corrosion Mitigation
Consultants**
Whitlock Dalrymple Poston &
Associates, Inc. (WDP)

Environmental Consultants
Ambient Labs, Inc.

**As-Built Laser
Documentation**
Quantapoint, Inc.

Non-Destructive Evaluation
GB Geotechnics Ltd. (GBG)

Cathodic Protection
Electro-Tech CP

Donor Representative
Preservation Design

Construction Manager
F. J. Sciame Construction LLC

Window Subcontractor
Torrence Windows

Replacement steel frame window walls at the third and fourth floors, 2019.

A major industrial development related to thermal coatings has occurred recently as well. Ten years ago, thermal coating technology had the promising potential of improving the thermal performance of non-thermally broken steel frames by reducing the potential for condensation, but progress had remained at the research and development level. They are now a reality and available through at least one major manufacturer of high performance architectural coatings. Although the texture and appearance of the thermal coatings available today is far from the desired finish for an exterior glazed assembly, ongoing research and development coupled with increased market demands will certainly lead to more suitable products.

Another promising development is the architectural application of low voltage translucent coatings for heated glass (similar to what has been used for years on car windshields, freezer doors at supermarkets and glass towel warmers). These are now being sold commercially, along with UL-approved architectural grade controllers that were not available ten years ago.

The combination of thermal coatings and low voltage heated glass has tremendous potential to improve the performance of Modern single-glazed steel frame assemblies that only ten years ago had to be replaced to improve the indoor environments. Once this potential is exposed, more commercially available products and systems are to be expected, just as happened with thermally broken steel frame systems. As with all new systems, it is important that these solutions are both monitored in the field and tested against current (and future) industry and preservation standards to confirm not only their performance, but also to assess their impact on historic assemblies.

NOTES

1 Angel Ayón and William B. Rose, "Glazing Upgrade at Frank Lloyd Wright's Solomon R. Guggenheim Museum," *paper presented at Building Enclosure Science & Technology (BEST 2) Conference, Portland, OR, 12–14 April, 2010.*

2 Angel Ayón and William B. Rose, "Reglazing Frank Lloyd Wright's Solomon R. Guggenheim Museum," *APT Bulletin* 42, no. 2/3, Special Issue on Modern Heritage (2011): pp. 59–60.

3 Ibid., pp. 61–64.

4 Downsizing the mechanical equipment in all likelihood translated into energy savings for the museum, although unfortunately no records were collected before and after the intervention to validate this assumption.

5 Wright to Hilla Ribay, 13 March 1944, Frank Lloyd Wright correspondence, Research Library at the Getty Research Institute for the History of Art and the Humanities.

Yale University Art Gallery

New Haven, Connecticut, USA
Louis Kahn, 1953

North and York Street (west) facades of the Yale University Art Gallery in New Haven, Connecticut, ca. 1953.

North (courtyard) facade shortly after construction, 1952.

Designed by American architect Louis Kahn (1901–1974), the Yale University Art Gallery continues the legacy of expanding and adding to existing buildings designed by other architects that defined Classical architecture for centuries. Yet, unlike the Classical tradition, where stylistic variations consistently deferred to universal harmony, Kahn's addition to Yale's 1928 Italianate neo-Gothic museum by Egerton Swartwout is both reverent and self-referential.

In plan, the addition is based on two rectangles of different size aligned next to each other. The larger one is aligned with the original museum on Chapel Street through a windowless wall on the south facade. A modular building-height window wall dominates the north and west facades. A sunken courtyard on the York Street front creates an enigmatic yet inviting buffer that magnifies the west window wall. The smaller rectangle defining the plan, which also has a windowless south wall, creates a setback that accommodates the main entrance through a platform located a few feet above the street level. In Kahn's carefully studied entrance sequence to the gallery, one's perception transitions from the size, weight and scale of the neo-Gothic stonework and its leaded-glass windows to the simpler scale and dimension of the brick units in the windowless walls, and from there to the reflective and translucent quality of the Modern steel frame window wall with IGUs and its linear framing.

York Street (west) facade and roofed-over west courtyard before rehabilitation, ca. 2003.

CONDITION PRIOR TO INTERVENTION

From the start, there was excessive thermal expansion and interior condensation occurring at the window walls.[1] Soon after the

Chapel Street main entrance after the window wall replacement, 2016.

Replacement thermally broken aluminum frame window wall at the Chapel Street main entrance, 2016.

building's opening in 1953, drip pans were installed along the bottom interior side of the window wall to collect the condensation. The main steel frames of the window walls that were originally embedded within the concrete floor structure were damaged due both to the lack of adequate tolerances for thermal expansion and from rust jacking. Perimeter seal failures at the original IGUs caused the window wall to become opaque as moisture migrated in between the glass panes.[2]

Over time, exhibition design and art conservation standards also evolved to value visual neutrality and stricter control of the hydrothermal environment. As the building program changed, various rehabilitation campaigns followed. Many spaces were altered to become more neutral, covering some of the original walls with gypsum board. The cylindrical stair was enclosed to create additional storage space. The exterior York Street court was roofed over to create a gallery.[3]

INTERVENTION

Eventually, the significant deterioration of the Kahn building and pressure to create a modern museum environment convinced Yale to move ahead with a major rehabilitation project, and in 2000 they hired Polshek Partnership (now Ennead Architects). The work performed focused on the integration of new museum-grade mechanical systems and replacement of the steel-framed window wall with a thermally broken aluminum window wall system. The output of the replacement fin-tube radiators along the

TIMELINE

1928	Original museum building designed by Egerton Swartwout is completed
1953	Louis Kahn's Yale University Art Gallery addition completed
1963	Art and architecture studios move into the top floor
1994	Yale commissions a technical assessment of the four-story building
1995–1998	Master planning for the Yale Campus Arts Area completed
1998–2000	Preliminary design study completed for the three-building complex
2003	Rehabilitation project by Polshek Partnership Architects (now Ennead Architects) begins
2006	Restoration work is completed

Restored west courtyard and replacement thermally broken aluminum frame window wall at the York Street (west) facade, 2016.

perimeter of the building was doubled. The light-controlling fine-woven textile drapes original to the building were replaced. The replacement IGUs included two low-e coatings—one on glass surface #2 to control heat gain and another on the #3 surface to retain the heat generated by the new fin-tube radiators. The extent of corrosion at the window wall steel frames required removal of the original frame and installation of a customized thermally broken aluminum window wall system. To accommodate differential movement within the new aluminum window wall system, new inter-story joints were inserted. The replacement system

Replacement thermally broken aluminum frame window wall at the north courtyard facade, 2016.

Detail view of replacement thermally broken aluminum frame window wall at the York Street (west) facade, 2016.

matched the sightlines of the original steel frame assembly but was slightly deeper. Only the slab edge cover required a new thermally improved fascia and was, therefore, reinstalled off the plane.[4] These nearly imperceptible changes have enhanced the window wall performance and have allowed the continued use of the building for its original purpose without compromising its design intent, historical appearance and significance within the university campus.

COMMENTS

The window wall replacement work at Louis Kahn's Yale University Art Gallery is a unique case study where, similar to the Guggenheim Museum (see page 210), the original steelframe system was deemed unsuitable to meet the environmental performance required by contemporary conservation standards for museum collections. Unlike the Guggenheim though, the Yale Art Gallery was not a listed or designated local landmark, yet the significance of the window wall was recognized and enhanced by the intervention team nonetheless. The replacement work re-established the prominence of the window wall within the original building design through a combination of removal of unsympathetic additions and retention of the original overall appearance and detailing. It reinstated the window wall as the linchpin to a seamless transition between the original neo-Gothic museum and its Modern addition along Chapel Street. The new matching replacement window wall also reinforced the relationship between the building and its

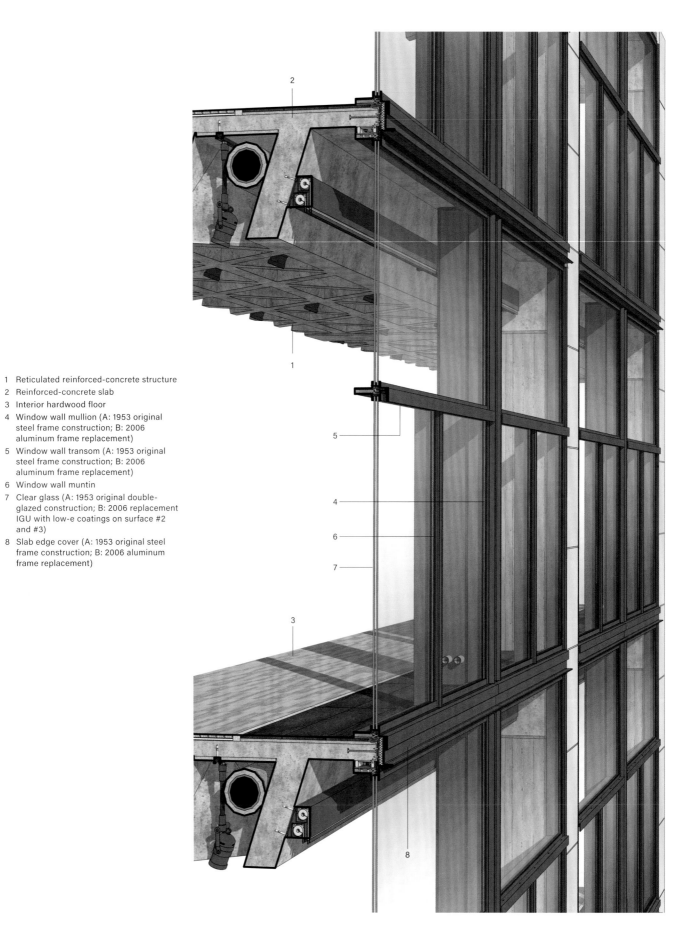

1 Reticulated reinforced-concrete structure
2 Reinforced-concrete slab
3 Interior hardwood floor
4 Window wall mullion (A: 1953 original steel frame construction; B: 2006 aluminum frame replacement)
5 Window wall transom (A: 1953 original steel frame construction; B: 2006 aluminum frame replacement)
6 Window wall muntin
7 Clear glass (A: 1953 original double-glazed construction; B: 2006 replacement IGU with low-e coatings on surface #2 and #3)
8 Slab edge cover (A: 1953 original steel frame construction; B: 2006 aluminum frame replacement)

Typical section at existing replacement aluminum frame window wall.

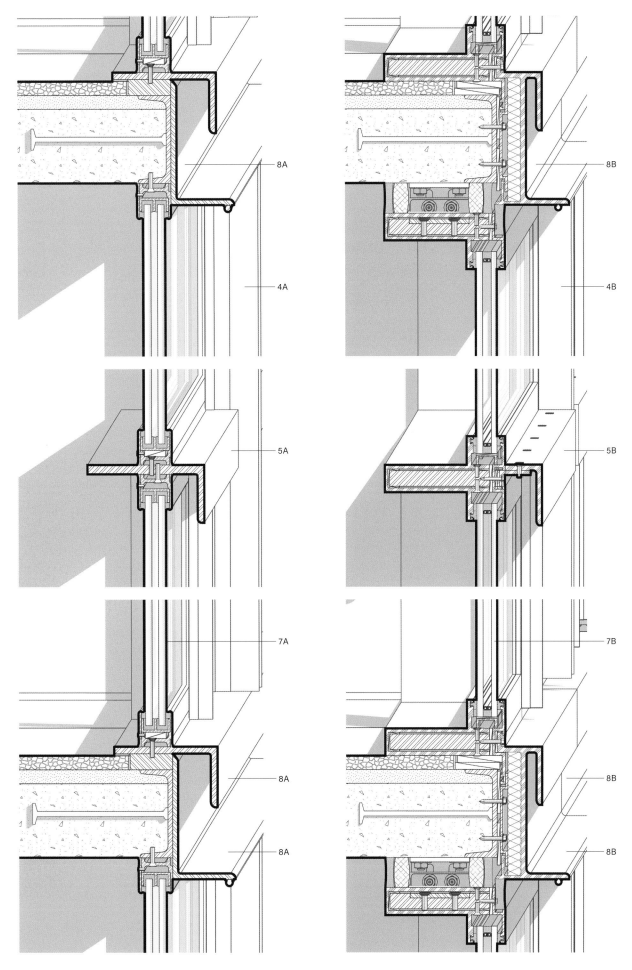

8A

4A

5A

7A

8A

8A

8B

4B

5B

7B

8B

8B

Typical before (left) and after (right) window wall details.

Replacement thermally broken aluminum frame window wall at the York Street (west) facade, 2016.

CREDITS

Yale University Art Gallery
(1953)
New Haven, Connecticut

ORIGINAL CONSTRUCTION

Architect
Louis Kahn

1998–2006 INTERVENTION

Architects
Polshek Partnership Architects
(now Ennead Architects)

Structural Engineer
Silman

MEP/FP
Altieri Sebor Wieber LLC
Consulting Engineers

Civil
BVH Integrated Services

Landscape
Towers|Golde Landscape
Architects and Site Planners
LLC

Lighting
Hefferan Partnership Lighting
Design

Acoustics
Shen Milsom & Wilke

Security
Ducibella Venter & Santore

Specifications
Robert Schwartz & Associates

Building and Life Safety Code
Hughes Associates, Inc.

Cost Estimating
Vermeulens Cost Consultants

Exterior Envelope
Simpson Gumpertz & Heger,
Inc.

Historic Preservation
Building Conservation
Associates, Inc.

Construction Manager
Dimeo Construction

Replacement thermally broken aluminum frame window wall at north courtyard facade, 2016.

Modern approach to site-making at both the adjacent sunken yard on York Street and the sculpture garden to the north.

What was found wanting in the renovation of Kahn's Yale University Art Gallery is the appearance of the double-coated IGUs, which is particularly noticeable where the replacement interior window treatments are deployed throughout the window wall. The presence of the low-e coating on the interior face of the exterior glass pane (surface #2) lends a purple-green hue to the glass that introduces a color gradation to a glazed assembly that was originally made of translucent clear glass. Unfortunately, even with the new mechanical system, the sheer expanse of the glazed enclosure made this subtle, but notable, design change a necessity. The double coating enhances the performance of the window wall, but it also increases its reflectivity and introduces polychromatic tones where there were none. This change in the visual properties of the glass is only reinforced by the off-white color of the interior window treatment. This is not only a departure from the original design intent and historic appearance of the window wall, it is also an unintended change that is regrettably both highly reflective and visually distracting.

NOTES

1 Lloyd L. DesBrisay, "Yale University Art Gallery: Louis I. Kahn: Challenges for the Rehabilitation of Modern Museum Buildings," *Journal of Architectural Conservation 13*, no. 2 (July 2007): p. 76.

2 Ibid., p. 77.

3 Ibid., pp. 74–75.

4 Ibid., pp. 80–82.

Conclusions

Summary of Intervention Strategies

The case studies in this publication highlight a variety of intervention strategies for the steel frame glazed enclosures of some of the most iconic Modern buildings in the US and Europe. While each has been included in one of main approach sections of this book (either restoration, rehabilitation or replacement), in reality, it is very common for more than one approach to be implemented on the same building or historic site. The nature of the materials and often experimental designs and previous histories of repair or disrepair all contribute to the complexity and uniqueness of each site's existing conditions, as do a range of other less tangible factors in the form of code regulations, energy-consciousness, or efforts to preserve historical and cultural heritage, among others. The decision-making process that ultimately leads to the implemented solutions at any given site is always unique, but by analyzing the resulting interventions across the case studies used in this book, a range of distilled subcategories was revealed which have been summarized below. While additional subcategories are possible beyond the limited scope of this book, we hope the reader may find these a useful reference.

The first intervention strategy summarized in this book relates to the restoration of 20th-century historic exterior glazed assemblies. Restoration is a reglazing intervention suitable for culturally significant buildings where the extent of surviving original historic fabric or the availability of reliable archival documentation merits addressing ongoing decay at the steel frames and retaining or replacing the original single-pane glass as needed. This intervention strategy can be implemented according to various subcategories that include the following:

FRAME REPAIR AND SINGLE-PANE GLASS
This approach involves repairing existing steel frames by cutting out and replacing in-kind missing or severely deteriorated frame and sash sections, glazing beads, hardware and other components, as well as replacing any related putty or perimeter sealants (joint sealers that test positive for hazardous materials should be abated and legally disposed of in accordance with applicable environmental regulations). Where original single-pane glass ex-

JK Building in Belo Horizonte, Minas Gerais, Brazil (Oscar Niemeyer, 1959). Corrosion at the original facade, which included large steel frame window walls with clear vision panels and spandrels made of translucent single-pane wire glass, 2008.

ists, it is retained or replaced in-kind as far as possible with similar single-pane glass. This subcategory is best represented by the case studies for the Villa Tugendhat in Brno, Czech Republic (Ludwig Mies van der Rohe, 1930), Halls 2 and 5 of the Zeche Zollverein complex in Essen, Germany (Schupp and Kremmer, 1932 and 1961), the staircase window wall at Viipuri Library in Vyborg, Russia (Alvar Aalto, 1935) and the Glass House in New Canaan, CT, USA (Phillip Johnson, 1949).

The restorative approach of this intervention strategy is appropriate for conditions where all or most of the original frame and glass is in place and where there is extensive surviving original historic fabric that allows for the selective replication of missing or severely deteriorated components. This approach is the gentlest and most conservative of all the available intervention strategies. It prioritizes retention of the historic appearance, fabric and design intent, and addresses safety, ahead of any enhanced environmental performance.

FRAME REPAIR AND SINGLE-PANE LAMINATED GLASS

This approach is very similar to the previous one, however, it focuses on improving glass impact safety and/or enhancing interior UV protection through the installation of single-pane laminated glass in lieu of original single-pane plate or other glass types. This subcategory is best represented by the case studies for the Hallidie Building in San Francisco, CA, USA (Willis Polk, 1918), Fallingwater in Mill Run, PA, USA (Frank Lloyd Wright, 1937), and the TWA Flight Center at JFK Airport in Queens, NY, USA (Eero Saarinen, 1962).

This intervention strategy is appropriate when replacement of the original single-pane glass is required to address either safety concerns such as a particularly large glass size, or a location over-

Corrosion and material decay (similar to what is shown here) is often due to a lack of maintenance. These and other conditions put many Modern steel frame glazed enclosures at risk of replacement due to safety concerns and the high cost of repairs, 2013.

head or within proximity of finished floors (as was the case at the Hallidie Building and the Convent of La Tourette), or to satisfy specific impact resistance requirements dictated by the building program or site constraints. It is also suitable for conditions where all or most of the original frame is in place but the original single-pane glass is not available due to ongoing decay or its removal during a previous replacement campaign. The benefit of this approach is that it protects the historic metalwork; however, depending on the glass selection, there is the risk that it may compromise the integrity and authenticity of the historic appearance. Like the previous approach, it prioritizes safety and UV protection (see Fallingwater case study) and does not focus on enhancing the building's environmental performance.

FRAME REPAIR AND SINGLE-PANE GLASS WITH SURFACE-MOUNTED FILM

This approach shares the same scope of work as the previous two regarding repair of the existing steel frames and localized in-kind replacement of missing or severely deteriorated components, but it also retains the original single-pane glass. It focuses on improving the impact safety of the glass and enhancing visible light protection and/or interior UV protection, sun and glare control through the installation of surface-mounted films. Past interventions from the case studies of Fallingwater in Mill Run, PA, USA (Frank Lloyd Wright, 1937), the Farnsworth House in Plano, IL, USA (Ludwig Mies van der Rohe, 1951), and the TWA Flight Center at JFK Airport in Queens, NY, USA (Eero Saarinen, 1962), have used this approach.

The surface-mounted films protect the single-pane glass, but their limited lifespan means their effectiveness is relatively short-lived before they start to peel away and detach (see previous interventions in the Fallingwater case study). Another disadvantage of this approach is the limited product sizes that are available in the cur-

399 Park Avenue in New York City (Carson Lundin and Shaw, 1961). Ongoing curtain wall over-clad, 2016.

rent market. While the film roll lengths are typically sufficient to accommodate existing glazing heights, each roll has a limited width which requires installations to have small overlaps or butt joints. This introduces discrete (but noticeable) vertical seams when installed on wide glass panes (see previous intervention in the Farnsworth House case study). Another drawback to some sun and glare control films (and often of UV protection films) is that they work by changing the visible light properties of the glass, often reducing visibility (see previous interventions in the TWA case study). This approach prioritizes safety and user comfort, and does not (like the previous two restoration approaches) focus on enhancing the building's environmental performance. It is not recommended as a long term solution, but it is a viable option to consider when a short-term intervention is necessary before a long-term repair program and budget are put into place.

The second intervention strategy summarized in this book relates to the rehabilitation of 20th-century historic exterior glazed assemblies. Rehabilitation is a reglazing intervention suitable for glazed assemblies with significant material loss at the metal frames or of the glass, and where in-kind replacement has been deemed unfeasible, ineffective or inappropriate, or where upgrading to meet contemporary environmental or structural performance requirements is required to facilitate ongoing or new programs. This intervention strategy can be implemented according to the following subcategories:

FRAME REPAIR AND INSULATED GLASS UNITS (IGUs)

This approach includes the steel frame repair scope outlined for the previous restoration subcategories, as well as removal of the original single-pane glass and installation of replacement insulated glass units (IGUs). This subcategory is represented by Halls 7 and 9 of the Zeche Zollverein complex case study in Essen, Germany (Schupp and Kremmer, 1932 and 1961).

The installation of new IGUs in restored frames usually means the old putty or glazing beads need to be removed, and new glazing beads designed to accommodate the new glazing. Given that an IGU is thicker and heavier than the original single-pane glass, the replacement glazing bead is typically shallow and high-strength in order to be able to retain the IGU in place. Industry standards relating to the required bite size, face and edge clearances must also be considered when designing IGU and glazing bead installations on existing frames. This approach allows Modern glazed enclosures with poor thermal performance to be adapted for new programs and contemporary energy conservation requirements. It improves environmental performance and may also improve safety (depending on the glass make-up). It is recommended as a long-term solution, but only after careful consideration of a number of factors, a primary one being whether the existing frames and supports have the capacity to accommodate the new loads imposed by the IGUs, which more than double the weight of the original single-pane glass. As evidenced by some of the case studies, it is advisable to coordinate the location of the replacement IGUs with the interior programs. Instead of implementing a building-wide wholesale glass replacement, consideration should be given to doing so only at strategic locations where the program requires enhanced performance (see the Zeche Zollverein case study).

FRAME REPAIR AND SINGLE-PANE GLASS PLUS SECONDARY GLAZING

This approach retains the original exterior glazed assembly, involving the same steel frame repair scope outlined for the previous restoration categories and the retention of the original single-pane glass. In addition, it includes the installation of secondary glazing. This subcategory is represented by the case studies for the Van Nelle Factory in Rotterdam, Netherlands (Brinkman & Van der Vlugt, 1931), Halls 6 and 10 of the Zeche Zollverein complex in Essen, Germany (Schupp and Kremmer, 1932 and 1961), and the Hardenberg House in Berlin, Germany (Paul Schwebes, 1956). Although not installed on a metal frame assembly, the intervention on the skylights of the Viipuri Library in Vyborg, Russia (Alvar Aalto, 1935) is also representative of this approach.

In the simplest version of this approach, the secondary glazing consists of an additional interior operable in-swing sash with single-pane glass located immediately adjacent to the original single-pane sash at the exterior glazed enclosure. When closed, this creates an air-tight cavity between both glazed assemblies while allowing for maintenance and cleaning when required (see the Hardenberg House and Viipuri Library case studies). In other cases, where it is feasible to replicate the exterior enclosure, the secondary glazing is installed on the exterior side of the original

1271 Avenue of the Americas in New York City (Wallace Harrison of Harrison, Abramovitz & Harris, 1959). Ongoing curtain wall replacement, 2018.

single-pane glass (see skylight work at Viipuri case study). This strategy is appropriate for situations where it has been determined that minimum enhancements of the exterior glazed enclosures are sufficient to extend their service life while retaining not only the original historic fabric but also most of the interior appearance. A more elaborate solution includes the installation of a new insulated interior partition on the inboard side of the exterior walls. The insulated partition also includes a new steel (or aluminum) frame assembly in front of the existing exterior openings. The secondary glazing includes in-swing sash with IGUs for cleaning and maintenance (see Halls 6 and 10 in the Zeche Zollverein case study).

This strategy is suitable for situations where the increase in the nominal thickness of the rehabilitated exterior walls (original wall plus adjacent interior insulated partition with supplemental glazing) is acceptable. One downside is that the proximity of the new supplemental glazing to the original exterior glazed assembly creates shadowing within the openings when viewed from the exterior. This effect has the potential to compromise the historic appearance and so it is important that this solution be strategically implemented to minimize any adverse impacts to the exterior.

An even more sophisticated variation of this solution includes placing the new insulated interior glazed partition a greater distance away from the original single-pane exterior glazed assembly, thereby creating a circulation space and retaining the character of the exterior glazed assembly from both the exterior and the interior (see the Van Nelle Factory case study). This strategy is suitable for situations needing a value-based approach that prioritizes authenticity and differentiation of materials and systems, as well as retention and appreciation of historic character over maximizing all available usable space.

Hotel Tivoli in Maputo (Carlos Veiga Pinto Carmelo, 1970). The facade represents an unusual combination of Modern steel frame window walls and the plasticity of the Brutalist-style exposed reinforced-concrete, 2009.

FRAME REPAIR WITH STRUCTURAL REINFORCE-MENT AND SINGLE-PANE GLASS

This approach also includes the repair scope outlined for the previous restoration categories regarding repair of the existing steel frames and localized in-kind replacement of missing or severely deteriorated components, as well as retention or replication of the original single-pane glass. The distinctive feature of this approach is that discrete supplemental structural reinforcements are provided to reinforce the exterior glazed enclosures against lateral loads and/or to resupport them where required to address deficiencies in the original design or construction. This subcategory is represented by the Hallidie Building in San Francisco, CA, USA (Willis Polk, 1918), and the De La Warr Pavilion in Bexhill-on-Sea, UK (Erich Mendelsohn and Serge Chermayeff, 1935).

It is critical that the size, location and configuration of the supplemental reinforcement is carefully designed to ensure that they are not unsightly and do not obliterate the prominence of the restored original features. Another critical issue to be taken into account relates to the coordination of the finishes applied on both the original frames and the supplemental reinforcement. There are situations where the same finish is applied to both as a way to blend one with the other (see the De La Warr Pavilion case study). In other cases, decisions have to be made as to whether or not the supplemental reinforcement should be differentiated from the original material (see the Hallidie Building case study).

The third intervention strategy summarized in this book is the replacement of 20th-century historic exterior glazed assemblies. Replacement is a reglazing intervention strategy where, despite the possible significance of a building and its original architectural character, the original steel frames and glazing are removed and replaced with new assemblies. This intervention strategy retains the original appearance as much as possible, but focuses primarily on improving environmental performance and safety. Replacement is the most invasive of the three interventions presented in this book, as it prioritizes safety, energy conservation and user comfort over conservation of the original historic fabric. This intervention strategy can be implemented according to the following subcategories:

NON-THERMALLY BROKEN STEEL FRAMES AND SINGLE-PANE GLASS

This approach involves the removal of severely deteriorated or poorly performing existing steel frames along with any associated glazing putty or perimeter sealants. New steel frames, sashes, glazing beads, hardware and other components as applicable are installed, as is new single-pane glass. Similar to other approaches mentioned above, if obsolete joint sealers test positive for hazardous materials, they should be abated and legally disposed of in accordance with applicable environmental regulations. This subcategory is best represented by the case studies for the Lever House in New York, NY, USA (SOM, 1952), and the S.R. Crown Hall in Chicago, IL, USA (Ludwig Mies van der Rohe, 1956).

This approach is suitable for significant Modern buildings where retention of the original appearance is of utmost importance. In-kind replacement, with a few concealed improvements not visible from the exterior, has been implemented in cases where extensive corrosion prevented the reuse of original steel frames (see the Lever House case study). It is advisable when pursuing this approach to consider whether or not replacement glass that is visible from only one side needs to remain as a single-pane or if it could be upgraded to an IGU (see the subcategories below). Most often, the replacement single-pane glass is visible from both the exterior and the interior, and will be thicker in dimension than the original, requiring a deeper and wider glazing pocket than the existing one to allow for the required bite size, face and edge clearances. To accommodate this, the existing glazing beads or glass stops may need to be replaced (see the S.R. Crown Hall case study).

This approach prioritizes retaining the historic Modern appearance over improving the thermal performance of the glazed enclosures. Its applicability is limited and dictated by a combination of factors that include the building's cultural significance, the conservation state of the existing exterior glazed assembly and the need (or lack thereof) to implement passive thermal improve-

ments mandated by local performance requirements without re-
sorting to costly and energy-consuming interior mechanical solutions (see the Convent of La Tourette case study). A variation of
this approach is represented by the replacement non-thermally
broken aluminum frames with single-pane glass that exists at the
Bauhaus workshop in Dessau-Roßlau, Germany (Walter Gropius,
1926) and at the Convent of La Tourette in Éveux, France (Le
Corbusier, 1960).

NON-THERMALLY BROKEN STEEL FRAMES AND INSULATED GLASS UNITS (IGUs)

This approach includes the same replacement scope outlined
for the previous subcategory, except that the glass replacement
utilizes IGUs instead of single-pane glass. This subcategory is
represented by the case studies for the Fagus Factory in Alfeld,
Germany (Walter Gropius and Adolf Meyer, 1912 and 1925), Sanatorium Zonnestraal in Hilversum, Netherlands (Jan Duiker, 1928 to
1931), Halls 7 and 9 of the Zeche Zollverein complex in Essen, Germany (Schupp and Kremmer, 1932 and 1961), as well as the penthouse and garage storefront at City of Refuge in Paris, France (Le
Corbusier and Pierre Jeanneret, 1933 and 1953).

This approach is suitable for cases where the existing frames
are in poor condition (see the Sanatorium Zonnestraal case study)
or do not match the historic configuration (see City of Refuge case
study) and where the space program can benefit from the enhanced environmental performance achieved through the use of
IGUs. One consideration, however, is that under extreme weather
conditions the poor thermal performance of the replacement
non-thermally broken frames will lead to seasonal condensation.
To avoid this condition and to ensure optimum environmental performance, new thermally broken frames may be an appropriate
installation, provided there is a budget that allows it (see the subcategories below). At locations with less demanding performance
requirements, such as entrances and circulation spaces, restoration of the original single-glazed enclosure should be considered instead of replacement (see the Fagus Factory case study).

THERMALLY BROKEN STEEL FRAMES AND INSULATED GLASS UNITS (IGUs)

This approach also includes the removal and replacement scope
outlined for the previous replacement categories, but calls for the
installation of replacement IGUs instead of single-pane glass. The
main difference is that the steel frames are thermally broken,
which minimizes thermal losses, thereby reducing energy consumption and resulting in optimum environmental performance.
This subcategory is represented by the case studies for the Bauhaus office and residential studio buildings in Dessau-Roßlau,
Germany (Walter Gropius, 1926), Hall 9 of the Zeche Zollverein

Grande Hotel of Ouro Preto in Ouro Preto, Minas Gerais, Brazil (Oscar Niemeyer, 1940). One of Niemeyer's first works, the original steel frame sliding windows have been maintained through the decades, 2008.

complex in Essen, Germany (Schupp and Kremmer, 1932 and 1961), and the Solomon R. Guggenheim Museum in New York, NY, USA (Frank Lloyd Wright, 1959).

This solution is suitable in cases where enhanced environmental performance is desired in addition to retention of the historic appearance, but there is not a mandate necessitating the retention of original historic fabric that might be in good or fair physical condition. The replacement steel frames for these assemblies can be made of custom hot-rolled components assembled through a proprietary system (see the north wing and residential studio windows in the Bauhaus Dessau case study), cold-formed hollow metal (see Hall 9 of Zeche Zollverein case study) or a combination of the two (see the Guggenheim Museum case study).

THERMALLY BROKEN ALUMINUM FRAMES AND INSULATED GLASS UNITS (IGUs)

This approach includes the removal and replacement scope outlined for the previous replacement categories and the installation of IGUs to replace single-pane glass. The main difference is that the thermally broken frames are made of aluminum extrusions, which also minimize thermal losses, reducing energy consumption and resulting in optimum environmental performance. This subcategory is best represented by the case study for the Yale University Art Gallery in New Haven, CT, USA (Louis Kahn, 1953).

The applicability and appropriateness of each aforementioned intervention subcategory, or combination thereof, as applied to a specific Modern building is a value-based decision to be made by stakeholders in consultation with heritage conservation and design professionals. The decision-making process should be led by a combination of factors that include, but are not limited to, a determination of the historic, architectural and cultural significance of the site, a hands-on assessment of the nature and condition of the historic and existing fabric, the ongoing or new program requirements, the environmental goals set out for the renovation, as well as the project budget, phasing, schedule and the available building construction labor, experience, materials and equipment.

The authors hope that the case studies presented in this book serve as valid references to guide alterations on both culturally significant and ordinary buildings with Modern glazed enclosures. The combination of documented facts and authors' comments included in the presentation and evaluation of each case study illustrates the evidence-based methodology required to determine the effect of the proposed interventions on the character and significance of Modern buildings. By identifying, describing, categorizing and evaluating the individual and specific interventions at each case study, the authors have outlined a set of criteria and decision-making options that can be considered for other existing buildings. Their hope is that readers will therefore find valuable information and sound references on the case studies presented, and that the findings summarized in the book will assist with evaluating and intervening on other Modern buildings with glazed assemblies. That aspiration will hopefully make a positive contribution towards the appropriate restoration, rehabilitation and replacement of Modern glazed enclosures and the resulting betterment of the built environment at large.

General Recommendations

All the US and European case studies presented in this publication are subject to heritage protections. As a result, they had to follow a particular set of historic preservation rules and procedures that typically focus on safeguarding original materiality and authenticity. Due to the often experimental character of 20th-century architecture and construction, current guidelines and charters, such as Articles 5 and 10 of the 2017 Madrid-New Delhi Document, emphasize the need to "acknowledge and manage pressures for change" and to "give consideration to environmental sustainability."[1] Thus, one of the most recurrent reasons leading to interventions on Modern steel frame glazed assemblies is the need to implement energy-efficient design solutions. Regardless of how stringent energy efficiency requirements are in the US and Europe,

The Ministry of Education and Health Building (Edifício Gustavo Capanema) in Rio de Janeiro (Lucio Costa, 1943). The north facade includes both *brise soleils* and operable aluminum louvers that provide shading and allow for natural ventilation through the building's steel frame window walls, 2004.

exceptions apply to both places regarding their applicability to historic buildings. For instance, in New York, San Francisco and other US cities, listing on the National Register of Historic Places allows exceptions for compliance with energy conservation requirements mandated by the local building codes.

Examples of projects presented in this publication taking advantage of this exception are the case studies for the Lever House facade replacement and the Hallidie Building facade restoration. There are similar provisions in the International Building Code (IBC) and in many European ordinances, such as the German Energy Saving Ordinance (EnEV 2014/2016) that allows for exceptions for listed buildings and for a 40% higher annual primary energy consumption compared to new buildings.[2] Although similar regulations apply on both sides of the Atlantic, this book's findings indicate that the European case studies are more likely to be driven by a desire to implement and improve energy conservation and energy efficiency. Those case studies are valuable references for feasible solutions aimed at improving energy efficiency and user comfort in both Europe and the US.

Another relevant distinction between intervention approaches in the US and Europe relates to differences in their building design and construction industries. In the US and in Europe, window and curtain wall constructions are very prevalent in both existing post-war buildings and new construction. Starting in the 1960s, both regions experienced a noticeable shift from steel to aluminum frame construction. In the US, however, this type of facade construction assembly is often combined with more extensive mechanical systems to provide interior heating, ventilation and air conditioning. This marks a significant difference from European approaches where, due to higher energy costs, the use of natural ventilation in combination with energy-efficient mechanical systems is dominant. In Europe, this has led to more sophisticated and individualized solutions that combine building components with building services. Listed buildings and related research projects (Bauhaus Dessau, for example) have often played a key role in developing unique solutions that are subsequently further developed and adapted for implementation in the building design and construction market. This may also be due to the fact that many European countries have a strong tradition of small- and medium-size design and construction companies experienced in the development and adaptation of state-of-the-art building technology and with a culture of craftsmen who are more willing to apply them in smaller, individual projects than larger companies are.

Another critical factor to be taken into account is that the complexity and heightened performance requirements demanded of buildings—and from the building envelope in particular—have given rise to the building envelope/building physics consultant in the US and in Europe as a stand-alone specialist. Similar to historic preservation/heritage conservation practitioners, building envelope consultants often bridge the traditional boundaries between the architectural, structural and mechanical professions and combine them with contemporary knowledge on building physics, environmental design, materials science and constructability. Since the late 20th century, this has resulted in a professional scenario in which the design intricacies of new high-performance glazed enclosures have become the realm of building envelope practitioners. In the US, they are often the first professionals architects and owners call upon to intervene on aging glazed assemblies. Preservation professionals have not yet developed the knowledge and experience required to reclaim their role in the glazed facade renovation market, which is primarily dominated by professionals from the new construction industry. In Europe, however, the nature of the design profession and construction industry is more collaborative. Architectural practices are accustomed to the historic character of their cities and are often specialized, allowing preservation and technical implementation to be more integrated.

All aspects mentioned above demand better education of all practitioners involved. Thus, the curricula of architecture and engineering schools need to require more training in the areas of integrated and holistic building design and historic preservation. More specialized knowledge is also required in the field of building envelope analysis and design. Changes in the training and continuing education of historic preservation, facade design and real estate practitioners, as well as that for building owners, managers and officials, are also needed. Design and preservation professionals and owners alike should be further educated and trained in how to identify common ground between sustainability, historic preservation and building design goals. Likewise, additional technical training for metal workers and construction companies should be encouraged so that the maintenance and repair of existing steel frame enclosures becomes part of their standard skill set, equally relevant as the manufacturing and installation of new assemblies is within their trade. These observations are confirmed by different surveys made in the context of facade-related research.[3]

The restoration, rehabilitation and replacement of Modern glazed enclosures will also benefit greatly from the development of a practical set of guidelines to steer interventions in more appropriate directions both conceptually and technically. This is a topic that may necessitate reevaluating established preservation principles regarding protection of historic materials and systems, in an effort to balance the retention of historic fabric and design intent with the needs arising from sustainability and energy conservation mandates, owners' requirements and users' comfort. For these efforts to be successful, it is crucial that steps are taken to document previous interventions on glazed enclosures and that they are then assessed on a case-by-case basis to analyze the specific strategies used to address ongoing decay or improve outdated performance. Similar to how the case studies have been researched and presented in this publication, this work will require documentation of metal frame glazed enclosure details before and after the interventions, along with a critical assessment of the suitability and appropriateness of the work performed. These assessments should be based on how the work succeeded in protecting the original historic materials and systems, and how well contemporary sustainability principles were balanced with retention of original historic fabric and/or design intent.

Documentation and critical assessment are the most obvious point of departure when looking for the collective benefits and lessons that can be retrieved from these sets of dispersed but meaningful past interventions. The aforementioned initial steps will assist in setting up an evidence-based framework to identify best practices that can be used to generate guidelines for intervening on Modern glazed enclosures. It is also critical to undertake further study of interventions on metal-framed enclosures made of other materials, such as aluminum, bronze, brass and stainless steel or plastic, which were also extensively used throughout postwar Modernism. Except for a few cases, Modern-era buildings with exterior glazed enclosures made of these materials are typically larger and taller, and their appearance tends to be aligned with the International Style aesthetic. Poor thermal performance and weather protection, coupled with material decay and a growing tendency to upgrade the overall exterior image, among other factors, place these buildings at greater risk of alterations than their older, steel frame counterparts. It is imperative that the reglazing interventions on Modern buildings with these types of exterior glazed assemblies be evaluated, so that their poor thermal performance (use of aluminum), questionable sustainability and durability (use of plastics), higher cost (for metals such as bronze, brass and stainless steel) or lower public appreciation (sometimes associated with aluminum) do not become an excuse for inappropriate alterations.

One of the lessons learned through the case studies outlined in this book is that "Reglazing Modernism" is not limited to simply three main intervention strategies as presented here. Beyond subjecting the exterior glazed enclosures to restoration, rehabilitation, replacement and their various subcategories, the task of reglazing Modernism often requires less direct interventions, such as modifying the building systems, altering programmed space usage and adjusting comfort requirements. Even though restoration is the least intrusive, and therefore most desirable (from a historic preservation standpoint), of the three main intervention categories presented, it cannot always accommodate the performance requirements mandated by the ongoing or new building use. In these cases, rehabilitation options are appropriate; however, they should be implemented only where demanded by the interior use requirements. At locations with less stringent thermal performance requirements, such as glazed entryways (see Sanatorium Zonnestraal case study), existing circulation spaces (see Fagus Factory case study), new circulation or non mission-critical spaces resulting from the installation of secondary glazing (see Van Nelle Factory and Zeche Zollverein case studies), a restoration approach should be considered for the historic glazed assembly. When the building's use cannot tolerate vulnerabilities in the thermal performance of the exterior glazed enclosures, replacement with thermally broken systems and high-performance IGUs is an option to be considered as a last resort. This is often the case for museums and galleries (see the Guggenheim Museum and Yale University Art Gallery case studies), restaurants (see the Zeche Zollverein case study), offices and dormitories (see the Bauhaus Dessau case study). Even when replacement—the most intrusive approach—is pursued, the extent, slenderness and limited sightlines of new exterior glazed enclosures will often still require supplemental support to be provided by the building systems, as well as ongoing management of comfort expectations for users. For cold weather regions, that means reducing the heating degree day (HDD) temperature used as the benchmark for determining when heating is required—typically 65° Fahrenheit (18° Celsius)—or setting the thermostats to interior temperature values lower than expected at a newly built facility. Another supplemental strategy to be considered is implementing use restrictions during the colder periods (see the Bauhaus Dessau and Zeche Zollverein case studies).

The interventions discussed in this book are intended as only a starting point on the way to a larger assembled inventory with a wider range of intervention options for Modern buildings. These, and the next sets of projects to be presented by the authors in future publication(s) should serve as a reference for outlining general strategies for interventions on culturally significant and vernacular Modern buildings worldwide. It has to be acknowledged that the US and European approaches in this research cannot easily be adapted to other climates and cultures, as relevant factors can vary significantly according to local conditions. In order to improve overall building performance, solutions must not only adapt to the local climate, but also to the local building and construction marketplaces. User behavior and cultural differences between regions must also become part of the process. For all of these reasons, further research is needed to document, develop and assess suitable approaches originating from other climates, cultures and continents.

Recommendations to the Glass and Fenestration Industry

Additional research, development and wider dissemination of current and emerging technologies is required in the glass and fenestration frame industries if we are to expand the range of building envelope solutions available for interventions on Modern glazed assemblies. Within the glass industry, one area warranting this kind of attention is that of vacuum glazing technology. This type of glazing technology utilizes the superior thermal properties of a vacuum, which is less conductive than air or inert gases such as argon or krypton.[4] Vacuum glazing has been available in the European market for some time, but has only become available in the US within the last couple of years. The technology allows for the production of slim IGUs that have better thermal performance than IGUs constructed with air or inert gas infill. These thinner IGUs are also an easier fit into the shallow profiles of a rehabilitated or replacement steel frame or window sash (see Sanatorium Zonnestraal case study). Along with vacuum glazing, the glass industry needs to foster further development towards increasing the use of spacers fabricated with warm edge technology. These IGU spacers are made of proprietary materials with very low thermal conductivity instead of the highly conductive metals that are commonly used currently such as aluminum extrusions or cold-formed stainless steel. As a result, warm edge spacers significantly enhance the thermal performance of IGUs (see Guggenheim Museum case study).[5]

Another relevant technology that requires further research and development in the glass industry is the use of electrically-heated glass for exterior glazing. This technology is based on the installation of a translucent tin oxide coating on float glass. The coated glass is then laminated and fitted with two thin copper buss bars at opposite sides that can be concealed within the frame. Once the system is turned on, a low-voltage electrical current is transmitted between the buss bars and the thermal resistance of the tin oxide coating heats up, emitting radiant heat uniformly across the glass surface. The increased surface temperature can help prevent condensation and reduce energy losses from the interior. This technology, which is similar to the defrosting electrical heating in the laminated glass of a car windshield, to the towel warmers often found in the hospitality industry and to the glass doors of store freezers, is safe and has a very low energy consumption. It can be used for the interior glass of an IGU or, more importantly, for reglazing interventions at Modern buildings where use of single-pane glass is more appropriate than IGUs and there are no concerns about condensation on the frame. In 2004, this solution was considered for the Guggenheim Museum in New York City for a large, segmented ground-floor window wall facing Fifth Avenue. It could not be implemented at the time because the 240V power density demanded by the glass size required testing and Underwriters' Laboratories (UL) certification, which the manufacturer had not yet procured since they had never been required to produce heated glass at such size. Both 110V and 240V UL-approved controllers are available in the US and Europe today.[6] This technology should be further explored for its potential use in reglazing interventions requiring thermal improvements of single-pane glass.

Lastly, there is a need to support historic glass manufacturing to facilitate the fabrication and use of historic glass types, such as plate or drawn glass, with their characteristic roller wave appearance. The expanded availability of this type of glass has a market, not only for Modern reglazing applications, but also in the renovation of historic buildings from other periods.

In the fenestration frame industry, additional research and development of thermal break materials is required. Stiffer materials with increased thermal performance are needed, particularly for use in thermally broken steel frame systems which typically have slim and shallow frame profiles that pose a challenge to replacement interventions. Further development of thermal break materials and frame installation systems during the last decade has created some innovative and suitable products.[7] Still, more competition and availability of such systems in the marketplace are needed to overcome increased project costs.

Another technology that deserves further research and development for implementation in non-thermally broken steel frame systems is heat tracing. This low-voltage solution is similar to what is used for snow melting systems installed on roofs or in sidewalks, which warm up as a result of the electrical resistance produced by an induced current. For non-thermally broken steel frame systems, heat tracing offers an opportunity to improve the thermal performance of the frames, preventing condensation and condensate freezing. They can be successfully installed at discreet locations where removal and replacement of the existing frames is not desired, or where replacement thermally broken frames have to be installed on, or adjacent to, existing construction with inferior thermal properties (see the Bauhaus Dessau case study).

Finally, thermal insulating coatings are another emerging technology that deserves additional research and development. Based on Aerogel and other highly insulating materials that came out of the nanotechnology industry, thermal coatings are gaining presence in the construction industry. Today, they are primarily used for industrial applications, such as providing thermal protection for pipes (with a thin coating instead of applied insulation). Due to the rough appearance of the sprayed-on installation, their architectural application is currently limited, but they are being increasingly used to prevent condensation on ferrous embedments within masonry and to control condensation on steel and concrete building components, such as relieving angles, lintels, cantilevered beams, canopies and balconies.[8] Like thermal breaks for steel windows, which were nonexistent 15 years ago (see Guggenheim case study), thermal coatings have tremendous potential for architectural applications on non-thermally broken steel frame systems once the required dry film thicknesses can be achieved with a smoother, more aesthetically acceptable finish. Increasing the reliability and affordability of the aforementioned high-performance systems and making them more available in the marketplace will be a positive step towards increasing the fenestration industry's response to the challenges posed by some of the reglazing Modernism interventions summarized in this publication.

NOTES

1   ICOMOS International Scientific Committee on Twentieth-Century Heritage (ISC20C), *Madrid-New Delhi Document: Approaches for the Conservation of Twentieth-Century Architectural Heritage*, 2017, accessed 29 December 2018, http://www.icomos-isc20c. org/pdf/madrid-new-delhi-document-2017.pdf.

2   In 2019, the German Federal Government plans to unite the rules that still run in parallel (Energy Saving Act (EnEG), Energy Saving Ordinance (EnEV) and Renewable Energies Heat Act (EEWärmeG)) in a new Gebäude-EnergieGesetz – BuildingEnergyAct (GEG) that will incorporate all requirements from the European legislation.

3   Ulrich Knaack and Uta Pottgiesser, eds., *efnMOBILE 2.0: Efficient Envelopes* (Delft: TU Delft Open, 2017), p. 192, accessed 29 December 2018, https://repository.uantwerpen. be/docman/irua/031a68/147549.pdf.

4   Pilkington, "Pilkington Spacia™," accessed 9 March 2019, https://www.pilkington.com/ en-gb/uk/products/product-categories/ thermal-insulation/pilkington-spacia.

5   Joe Erb, "Defining warm edge in the commercial market," *Glass Magazine*, accessed 9 March 2019, https://glassmagazine.com/ article/commercial/defining-warm-edge-commercial-market.

6   See: Ayón and Rose, "Reglazing Frank Lloyd Wright's Solomon R. Guggenheim Museum," pp. 59–60; Sweets Construction, "Thermique™ Heated Glass for Architectural Windows," http://sweets.construction.com/swts_content_files/56074/5888_088100-THE_09.pdf; and HTG, "Heated Glass," http://htgglass.com/ en/products/75/heated-glass, both accessed 9 March 2019.

7   See steel profiles for facade systems at MHB's website: https://www.mhb.eu.

8   Tnemec, "Experience Thermal Break – Redefined," accessed 9 March 2019, http:// thermalbreak.tnemec.com.

Glossary

Annealed glass

Float glass that has not been heat-treated. See *Float glass*.

Awning window

A window with sash hinged at the top (header) of the frame.

Butt-glazing

Glazing installation with top and bottom frame support, where the vertical glass edges are without structural supporting mullions.[1]

Butt-joint

Joint made by parts meeting end-to-end without overlap. See *Butt-glazing*.

Casement window

A window with an in- or out-swing sash hinged at the side of the frame (jamb).

Curtain wall

A non-load-bearing exterior wall assembly, which is hung on the exterior of the building, generally constructed with metal frames and glass, metal and/or thin stone panels. Curtain walls are usually installed spanning floor-to-floor and do not carry floor or roof loads.[2] The wind and gravity loads of the curtain wall are typically transferred at the frame attachment to the floor line of the building structure.[3]

Double-glazed; double-pane glazing

In general, any use of two parallel lites of glass, separated by an air space to prevent heat loss or to improve sound attenuation. See *Insulating glass unit (IGU)*.

Drawn glass

Common until the 1970s, drawn glass was produced by drawing out and then cooling semi-molten glass through a series of metal rollers set up on automatic machinery. The process causes slight irregularities on the glass surface (referred to as "roller waves"), which distort light, causing non-specular reflections that are particularly apparent when the glass surface is viewed tangentially.

Edge clearance
Nominal spacing between the edge of the glass product and the bottom of the glazing pocket.[4]

Face clearance
Nominal spacing between the face of the glass product and the side of the glazing pocket.

Float glass
The float glass manufacturing process, first invented by Sir Alastair Pilkington in 1952, is used to make flat glass without the "roller wave" surface irregularities characteristic of drawn glass. It makes possible the manufacture of clear, tinted and coated glass for architectural and industrial applications.[5] The manufacturing method revolutionized the flat glass industry once it was put into practice in 1959. During the float glass process, molten glass from the furnace flows by gravity and displacement onto a bath of molten tin, forming a continuous ribbon. The glass ribbon is then pulled or drawn through the tin bath. Once it exits onto guiding rollers it is run through an annealing lehr where it is cooled under controlled conditions. When it emerges near room temperature, it is flat with virtually parallel surfaces and has a smooth, even finish. Automatic cutters are typically used to trim the edges and cut across the width of the moving ribbon. This creates sizes, which can be shipped or handled for further processing. Float glass that has not been heat-treated is referred to as annealed glass. Nearly all flat glass produced worldwide is manufactured using this process.[6] Use of float glass as replacement for original drawn or plate glass may require some compromises to be considered. Aside from the visual differences between the products, there are limitations to the sizes available for today's float glass sheets which may pose additional challenges for reglazing interventions at Modern buildings that require large sizes.

Fully tempered (FT) glass
Glass that has been heat-treated to increase strength, sometimes called "toughened glass" outside North America.[7] Glass with fully tempered surfaces is typically four times stronger than annealed glass and twice as strong as heat-strengthened glass of the same thickness, size and type. In the event that fully tempered glass is broken, it will break into fairly small pieces, reducing the chance of injury.[8] In order to meet the ASTM C1048 standard for fully tempered (FT) flat glass (EN12150 in Europe), the glass must have either a minimum surface compression of not less than 10,000 psi (69 MPa), or an edge compression of not less than 9700 psi (67 MPa), or meet other outlined government-mandated safety glazing requirements.[9]

Glass bite

The dimension by which the framing system overlaps the edge of the glazing infill.[10]

Glass stop

See *Glazing bead*.

Glazed assembly

See *Glazing*.

Glazing

Used as a verb, glazing refers to the process of installing glass (and sometimes other materials) into a prepared opening (with or without frames). Used as a noun, glazing refers to the transparent infill material (such as glass) used for windows, curtain walls, window walls, etc. The term glazing is also used as a noun to refer generically to glazed assemblies such as windows, doors, curtain walls, window walls or unit skylights.[11]

Glazing bar

See *Muntin*.

Glazing bead

A strip surrounding the edge of the glass in a window or door, which holds the glass in place.[12] Also referred to in this book as a *glass stop* when made of a solid steel bar.

Glazing pocket

Cavity within a frame where the glass is inserted, securely retained and provided with a given glass bite, edge clearance and side clearance.

Heat-soaked glass

The heat-soaking process is a method of reducing the incidence of spontaneous breakage in tempered glass caused by nickel sulphide (NiS) inclusion.[13]

Heat-strengthened (HS) glass

Glass that has been heat-treated to have a surface compression between 3500 and 7500 psi (24 to 52 MPa) and to meet the ASTM C1048 standards for heat-strengthened (HS) glass (EN1863 and EN12150 in Europe). HS glass is not a safety glazing material.[14]

Hopper window

A window with the sash hinged at the bottom of the frame, typically swinging inwards.

Insulated glass unit (IGU)

A type of glazing designed to reduce heat transfer through glass, where two or more panes are sealed together enclosing a hermetically sealed cavity. The panes are held apart by a perimeter spacer containing a desiccant to keep the enclosed cavity free of visible moisture. The entire perimeter of the assembly is sealed. The most commonly used edge construction contains a metallic spacer of roll-formed aluminum, stainless steel, coated steel or galvanized steel. It is sealed with a single seal of polysulfide, polyurethane or hot-melt butyl, or with a dual seal consisting of a primary seal of polyisobutylene and a secondary seal of silicone, polysulfide or polyurethane. Thermal performance of IGUs is enhanced by using solar-control substrates and coated glass (low-emissivity or reflective), coated polyester-suspended films, insulating inert gases (e.g. argon, krypton or xenon) and warm edge technology products. IGUs can be used to reduce initial heating and cooling equipment costs and ongoing operating costs, and can help reduce sound transmission as well.

Laminated glass

Two or more sheets of glass permanently bonded together with one or more interlayers.[15]

Lock-strip (or zipper gasket) glazing system

A two-part neoprene gasket assembly consisting of a main gasket body used to capture and support the glazed infill, and a lock-strip or "zipper" that is inserted into a receiving channel on the main gasket body to provide compression between the main gasket body, the glazing infill and the framing.[16]

Low-emissivity (low-e) coating

A coating applied to one or more glass surfaces in an IGU in order to alter the light transmission, reflection and/or absorption characteristics of different parts of the solar spectrum. Low-e coatings are installed using either the pyrolytic or the sputtered method. Pyrolytic low-e coatings, often called "hard coats", contain a conductive oxide infrared reflecting layer. They are usually tin oxide-based and are typically deposited during the float glass manufacturing process. Sputtered low-e coatings, often called "soft coats", contain multiple layers of metals, each with a different purpose (e.g. improving thermal or solar performance, modifying light transmission, reflectance, durability, etc.).

Low-iron glass

A type of high-clarity glass made from silica with very low amounts of iron. The low iron level removes the greenish-blue tint typically seen in large thicker sizes of glass.[17]

Mullion

Typically, a vertical member that supports panels, glass, sash or other sections of a glazed assembly.[18] Mullions generally separate two adjacent fixed panes, windows or doors.

Muntin

Muntins divide a single window sash, casement or glazed door into a grid system of small panes of glass, called lites. Also called glazing bars (UK).

Plate glass

Product first made by pouring molten glass on to a table and rolling it until flat, then grinding and polishing it into a plate. Advancements in the process led to feeding the molten glass though continuous rollers, grinders and polishers. Plate sheet glass is no longer produced commercially in the US or most European countries.[19]

Ribbon window

A series of windows set side-by-side to form a continuous band horizontally across a facade.[20]

Safety glass

Fully tempered or laminated glass commonly used for safety purposes. Tempered glass limits the risk of injury by fracturing into small fragments. Laminated glass limits the risk of injury by bonding the glass to PVB or SentryGlas® ionoplast interlayer(s), which hold the shattered glass fragments in place when a pane is broken. In the US, safety glazing must be identified with an indelible label on the glass indicating its conformance to published standards for public safety. Wired glass typically does not meet CPSC or European requirements for safety glazing.[21]

Sheet glass

Inexpensive glass produced in the early part of the 20th century by drawing a glass ribbon vertically out of a molten glass pool. Distortion in sheet glass is common due to the differences in viscosity of the molten glass. The later plate glass process was developed in an effort to eliminate this problem and produce relatively distortion-free glass for use in coach windows and mirrors. Sheet glass is no longer commercially produced in the US and Europe.[23]

Sightline
The line along the perimeter of glazing infill corresponding to the top edge of stationary and removable stops, or to the edge of the glazing sealant line.[22]

Single-glazed; single-pane glazing
Glazed assembly using monolithic glass of any type.

Spandrel
The panel(s) of a curtain wall located between vision areas of window walls, curtain walls, etc., which conceal(s) structural columns, floors and shear walls.[24]

Spectrally selective glazing
A type of glazing that reduces solar gain while still providing daylighting. Spectrally selective glazing has low solar heat gain coefficients (SHGC), high visible light transmittance (VT) and usually low U-values.[25]

Tempered glass
See *Fully tempered glass*.

Thermal break
A thermal break or thermal barrier is an element of low thermal conductivity placed in an assembly to reduce or prevent the flow of thermal energy between conductive materials.[26] In the fenestration industry, thermal breaks are usually made of polyamide, polyurethane or other low-conductive materials that are mechanically locked in place to reduce the thermal conductivity of aluminum or steel frames.

Thermally broken
A glazed assembly constructed with thermal breaks.

Toughened glass
See Fully tempered glass.

Transom; transom bar
Transverse horizontal bar or crosspiece separating a door from a fixed or operable unit above. The glazing unit above this crosspiece is often referred to as a *transom* or *transom window*.[27]

Vision glass
The section of a window wall or curtain wall glazed with translucent glass affording both daylighting and exterior views.

Warm edge

A type of IGU spacer made of extruded butyl materials, foam rubber-based materials, formed plastics or metal strip-based products, often with desiccant included as a component. Warm edge spacers offer lower thermal conductance than traditional aluminum spacers. Warm edge IGUs typically offer higher resistance to condensation and an incremental improvement in window energy performance.[28]

Window

An operable or fixed assembly installed in an opening within an exterior wall to admit light or air, usually framed and glazed.

Window wall

A non-load-bearing fenestration system comprised of a combination of assemblies and composite units, installed spanning from the top of a floor slab to the underside of the floor slab above and usually including vision panels (transparent or translucent) and/or non-vision panels (opaque glass or metal).

Wired glass

Glass made by feeding a welded wire net into the molten glass just before it enters the roller. In the event of breakage, the wire holds the glass lite in place.[29] Wired glass is primarily used in fire-rated products and it has been increasingly replaced by laminated glass as fire codes have been updated.

NOTES

1   *GANA Glazing Manual* (Topeka, KS: Glass Association of North America, 2004), p. 94.
2   For more information on curtain wall design requirements, see AAMA MCWM-1-89 Metal Curtain Wall Manual (Illinois: American Architectural Manufacturers Association, 1989).
3   Nik Vigener, P E and Mark A. Brown, "Building Envelope Design Guide—Curtain Walls," National Institute of Building Sciences, last modified 10 May 2016, accessed 11 March 2019, https://www.wbdg.org/design/env_fenestration_cw.php.
4   *GANA Glazing Manual*, p. 96.
5   Eurotherm, "Flat Glass Manufacturing," accessed 7 January 2019, http://www.eurotherm.com/flat-glass.
6   Glenny Glass Company, "History of Flat Glass Production," accessed 31 March 2017, http://www.glennyglass.com/history_of_flat_glass_production.htm.
7   *GANA Glazing Manual*, p. 97.
8   Viracon, "Heat Treatment," accessed 28 January 2019, http://www.viracon.com/page/heat-treatment.
9   *C 1048 Standard Specification for Heat-Treated Flat Glass—Kind HS, Kind FT Coated and Uncoated Glass* (West Conshohocken, PA: ASTM International, 2007), p. 4, accessed 1 March 2019, http://www.mdglass.net/wp-content/uploads/2014/01/ASTM-Tempered-Specification.pdf.
10 *GANA Glazing Manual*, p. 93.
11 Window & Door Manufacturers Association, "The Window Glossary," accessed 7 January 2019, http://www.wdma.com/?page=TheWindowGlossary; *GANA Glazing Manual*, p. 97.
12 *GANA Glazing Manual*, p. 97.
13 Protemp Glass Inc, "Heat Soaked Glass," accessed 7 January 2019, www.protempglass.com/heat-soaked-glass.html.
14 *GANA Glazing Manual*, p. 98.
15 Ibid.
16 Direct excerpt from: Chamberlin Roofing and Waterproofing, "Lock-strip Gasket Replacement & Remediation Strategies," accessed 6 January 2019, https://www.chamberlinltd.com/articles/lockstrip-gasket-replacement-remediation-strategies/.
17 Wikipedia, "Low-iron glass," accessed 7 January 2019, https://en.wikipedia.org/wiki/Low-iron_glass.
18 *GANA Glazing Manual*, p. 99.
19 Glenny Glass Company, "Insulating Glass," accessed 31 March 2017, http://www.glennyglass.com/history_of_flat_glass_production.htm.
20 Merriam-Webster, "Ribbon windows," accessed 7 January 2109, http://www.merriam-webster.com/dictionary/ribbon%20windows.
21 Vigener and Brown.
22 *GANA Glazing Manual*, p. 102.
23 Glenny Glass Company, "History of Flat Glass Production," accessed 31 March 2017, http://www.glennyglass.com/history_of_flat_glass_production.htm.
24 *GANA Glazing Manual*, p. 102.
25 Building Science Corporation, Glossary, accessed 7 January 2019, https://buildingscience.com/glossary/spectrally-selective-glazing.
26 Wikipedia, "Thermal break," accessed 7 January 2019, https://en.wikipedia.org/wiki/Thermal_break.
27 Wikipedia, "Transom (architectural)," accessed 7 January 2019, https://en.wikipedia.org/wiki/Transom_(architectural).
28 Direct excerpt from: National Glass Association, "Industry Resources: Glossary," accessed 6 January 2019, https://www.glass.org/industry-resources-glossary.html.
29 Direct excerpt from: National Glass Association, "Industry Resources: Glossary," accessed 6 January 2019, https://www.glass.org/industry-resources-glossary.html.

About the Authors

Angel Ayón, AIA, LEED AP, NCARB is the founder and Principal of AYON Studio Architecture · Preservation, P.C. (AYON Studio) in New York City, where he provides comprehensive professional services in the fields of architecture and historic preservation. Ayón has more than 24 years of experience of learning from, advocating for and ultimately conceiving and overseeing conservation efforts to save and secure our built heritage as a cultural asset for current and future generations. His experience with Modern architecture includes the rehabilitation and exterior enhancement of Frank Lloyd Wright's Solomon R. Guggenheim Museum for which he was the project architect between 2004 and 2008 (prior to founding AYON Studio). Among other organizations, he is a member of DOCOMOMO US and its local chapter DOCOMOMO New York/Tri-State, the Historic Preservation Committee of The Municipal Art Society of New York, the Board of Advisors of the Historic Districts Council, the Board of Trustees of the Preservation League of New York State and Save Harlem Now! He was awarded the 2015 James Marston Fitch Mid-Career Fellowship to undertake the research presented in this book. Ayón holds both a professional degree in Architecture (1995) and a MSc in Conservation and Rehabilitation of the Built Heritage (1998) from the Higher Polytechnic Institute José Antonio Echeverría in his native Havana, Cuba, and a Post-graduate Certificate in Conservation of Historic Buildings and Archaeological Sites (2002) from Columbia University in New York. For more about the author and his other publications, visit: www.ayonstudio.com/about1.

Uta Pottgiesser, PhD is Chair of Heritage & Technology at TU Delft in the Netherlands (Faculty of Architecture and the Built Environment) and is also Professor of Building Construction and Materials at TH OWL UAS in Germany (Detmold School of Architecture and Interior Architecture), where she served as vice-president (2006–2011) and dean (2012–2016). From 2016–2019 she was appointed Professor of Interior Architecture at the University of Antwerp in Belgium (Faculty of Design Sciences). She has more than 27 years of experience as a practicing architect and research scientist concerned with the protection, reuse and improvement of the built heritage and environment. Among other organizations, she is a member and vice-chair of DOCOMOMO Germany and is

chair of the DOCOMOMO International Specialist Committee of Technology (ISC/T). As a member of the European Facade Network (efn) and co-founder of the international Master of Engineering in Integrated Design (MID) program at TH OWL UAS, she researches, teaches and lectures internationally and continues to be a reviewer and (co-)author of several international journals and publications with a focus on construction and heritage topics, and is an active jury member in architectural competitions and PhD commissions. Pottgiesser holds a professional degree as an Architect (1993). She received her Diploma in Architecture from TU Berlin (1991) and obtained her PhD (Dr.-Ing.) from TU Dresden (2002), both in Germany, on the topic of "Multi-layered Glass Constructions. Energy and Construction." For more about the author, visit www.tudelft.nl/en/staff/u.pottgiesser/ and www.th-owl.de/fb1/en/fachbereich/personen.html.

Nathaniel Richards, LEED AP is a Senior Project Manager for JRM Construction Management and General Contracting. Richards has worked in the architectural design and construction management industry for more than 13 years, with a focus in the curtain wall and building envelope sectors for the past ten years. He has held instrumental roles in the engineering and construction of several of the tallest, most elaborate structures worldwide, including Lotte Super Tower in Seoul, Korea, Pacific Gate in San Diego, California, Comcast Tower II in Philadelphia, Pennsylvania and recently the New York Wheel in New York City. Throughout his career he has unraveled each project's complexity using the latest 3D software and analysis tools to illustrate, plan and manage the sequencing and logistics required to undertake challenging areas of each building. Richards obtained a BA in Architecture from Roger Williams University in Bristol, Rhode Island, US in 2006.

Illustration Credits

S. R. Crown Hall

1:       HB-18506-E2, Chicago History Museum, Hedrich-Blessing Collection.

2:       HB-18506-X, Chicago History Museum, Hedrich-Blessing Collection.

3:       HB-18506-Z4A, Chicago History Museum, Hedrich-Blessing Collection.

4:       HB-18506-S3, Chicago History Museum, Hedrich-Blessing Collection.

5:       HB-18506-X4, Chicago History Museum, Hedrich-Blessing Collection.

6-8/
10-12:   Photo by Angel Ayón.

9:       Photo by Peter J. Sieger, courtesy of Peter J. Sieger Architectural Photography.

Convent of La Tourette

1:       Photo by Pierre Varga, FOTO:FORT-EPAN / Varga, Pierre.

2:       Photo by p2cl, CC BY-SA 2.0, https://flic.kr/p/28dAKn.

3:       Photo by p2cl, CC BY-SA 2.0, https://flic.kr/p/28isX9.

4:       Photo by Fred Romero, CC BY 2.0, https://flic.kr/p/JaeeKA.

5-6:   Photo by Vanessa Fernandez.

7:       Photo by Aurelien Guichard, CC BY-SA 2.0, https://flic.kr/p/a8zNV8.

8:       Photo by p2cl, CC BY-SA 2.0, https://flic.kr/p/28dHPK.

9-12:   Photo by Uta Pottgiesser.

13:      Photo by Camster, own work, CC BY-SA 3.0, https://commons.wikimedia.org/w/index.php?curid=33285571.

Fagus Factory

1:       Photo courtesy of Nadine Gebaur/Fagus-GreCon.

2:       Photo by Edmund Lill:Klauss Lill, Courtesy the Bauhaus-Archiv Berlin.

3-5:   Photo courtesy of Nadine Gebaur/Fagus-GreCon.

6-15:   Photo by Angel Ayón.

Sanatorium Zonnestraal

1:       Photo by Unknown, courtesy of Het Nieuwe Instituut, no known copyright restrictions.

2:       Photo by Ronald Zoetbrood, courtesy of www.erzed.nl/enzonnestraal.html.

3:       Photo by Rijksdienst voor het Cultureel Erfgoed, CC BY-SA 4.0, https://commons.wikimedia.org/w/index.php?curid=24074512.

4:       Photo by Rijksdienst voor het Cultureel Erfgoed, CC BY-SA 4.0, https://commons.wikimedia.org/w/index.php?curid=23588018.

5:       Photo by Rijksdienst voor het Cultureel Erfgoed, CC BY-SA 4.0, https://commons.wikimedia.org/w/index.php?curid=23588001.

6:       Photo by Rijksdienst voor het Cultureel Erfgoed, CC BY-SA 4.0, https://commons.wikimedia.org/w/index.php?curid=23588007.

7-14:   Photo by Angel Ayón.

City of Refuge

1/3-4:   © F.L.C. / ADAGP, Paris / Artist Rights Society (ARS), New York, 2019.

2:       Photo by Pierre Yves-Petit (Yvon).

5-6:   Photo by Brian Brice Taylor.

7-16:   Photo by Angel Ayón.

Bauhaus Dessau

1:       Photo by Klaus Hertig, Stiftung Bauhaus Dessau (Besitz Scan), I 36041/1-2, © (Consemüller, Erich) Consemüller, Stephan (property of Original Vintage Print).

2:       Photo by Atlantic Photo-Co. Berlin, Bauhaus-Archiv Berlin.

3:       Photo by Lucia Moholy, ©2019 Artists Rights Society (ARS), New York / VG Bild-Kunst, Bonn.

4:       Photo by Unknown, Stiftung Bauhaus Dessau, I 17150 F.

5-15:   Photo by Angel Ayón.

Solomon R. Guggenheim Museum

1:       Photo © Ezra Stoller/Esto.

2-3:   Photo by William B. Short, courtesy of the Solomon R. Guggenheim Foundation.

4:       Exterior of the Solomon R. Guggenheim Museum, New York. Photo: David Heald © SRGF, NY (SRGM2016_ph149).

5-8/11: Photo by Angel Ayón.

9:       Café 3, Solomon R. Guggenheim Museum, New York. Photo: David Heald © SRGF, NY (Café 3 Monitor Space_ph03).

10:      Café 3, Solomon R. Guggenheim Museum, New York. Photo: Kristpoher McKay © SRGF, NY (SRGM2012_Tour_ph10).

Yale University Art Gallery

1-2:   Photo courtesy of the Yale University Art Gallery Archives.

3:       Photo courtesy of Ennead Architects (formerly Polshek Partnership).

4-10:   Photo by Angel Ayón.

Conclusion

1-2/5-6: Photo by Uta Pottgiesser.

3-4:   Photo by Angel Ayón.

7:       Photo by Marcos Leite Almeida.

Research Funding
The James Marston Fitch Charitable Foundation

Peer Review Comments
Nina Rappaport, New York
Pamela Jerome, New York
Laura Boynton, New York

Layout, Cover Design and Typesetting
Miriam Bussmann, Berlin

Editorial Supervision and Project Management
Henriette Mueller-Stahl, Berlin

Comprehensive Editing
Laura Boynton, New York

Copy Editing
Rosa Ainley, Strood, Kent

Production
Bettina Chang, Berlin

Lithography
Bildpunkt Druckvorstufen GmbH, Berlin

Paper
Magno Volume, 135 g/m^2

Printing
optimal media GmbH, Röbel/Müritz

The publication was made possible by the kind support of

Steel windows, doors and facades since 1938

MHB bv

OWL, University of Applied Sciences and Arts

Tnemec Company, Inc.

Library of Congress Control Number: 2019936745

Bibliographic information published by the German National Library
The German National Library lists this publication in the Deutsche Nationalbibliografie; detailed bibliographic data are available on the Internet at http://dnb.dnb.de.

ISBN 978-3-0356-1845-7
German Print-ISBN 978-3-0356-1847-1

© 2019 Birkhäuser Verlag GmbH, Basel
P.O. Box 44, 4009 Basel, Switzerland
Part of Walter de Gruyter GmbH, Berlin/Boston

9 8 7 6 5 4 3 2 1

www.birkhauser.com